The Physics of VLSI
(Xerox, Palo Alto, 1984)

AIP Conference Proceedings
Series Editor: Hugh C. Wolfe
Number 122

The Physics of VLSI
(Xerox, Palo Alto, 1984)

Edited by
John C. Knights
Xerox Palo Alto Research Center

American Institute of Physics
New York   1984

Copying fees: The code at the bottom of the first page of each article in this volume gives the fee for each copy of the article made beyond the free copying permitted under the 1978 US Copyright Law. (See also the statement following "Copyright" below.) This fee can be paid to the American Institute of Physics through the Copyright Clearance Center, Inc., Box 765, Schenectady, N.Y. 12301.

Copyright © 1984 American Institute of Physics

Individual readers of this volume and non-profit libraries, acting for them, are permitted to make fair use of the material in it, such as copying an article for use in teaching or research. Permission is granted to quote from this volume in scientific work with the customary acknowledgment of the source. To reprint a figure, table or other excerpt requires the consent of one of the original authors and notification to AIP. Republication or systematic or multiple reproduction of any material in this volume is permitted only under license from AIP. Address inquiries to Series Editor, AIP Conference Proceedings, AIP, 335 E. 45th St., New York, N. Y. 10017.

L.C. Catalog Card No. 84-72729
ISBN 0-88318-321-8
DOE CONF- 8408104

PREFACE

The International Conference on the Physics of VLSI (ICPVLSI) was a satellite conference to the 17th International Conference on the Physics of Semiconductors (ICPS) held in San Francisco, California, August 6-10, 1984. The ICPVLSI was organized by the committee listed below at the request of the ICPS Committee which acted on behalf of the IUPAP Semiconductor Commission which has sought to include bridges from the science to the technology of semiconductor physics in ICPS series. In addition to this satellite, two Symposia were organized within the main conference entitled: "Frontiers of Semiconductor Materials and Devices" and "Frontiers of Semiconductor Characterization". These are reported in the 17th ICPS Proceedings published by Springer-Verlag New York. The sponsors of the conference are listed on the next page.

The conference was held August 1-3, 1984, at the Xerox Palo Alto Research Center Auditorium in Palo Alto, California with 180 attending. These proceedings follow the organization of the conference which is presented in the Table of Contents. The subject matter of the conference spanned a wide range of topics of current importance to VLSI including: Oxides and Insulators, Surfaces and Interfaces, Interconnections and Silicides, Lithography, Reactive Ion Etching, Defects in Semiconductors, Physics of Devices, Materials Processing, Materials Modelling, Design Automation and Device Layout, and III-V Compounds.

ICPVLSI Committee:

Co-chairmen	Prof. J. D. Meindl	Stanford U
	Dr. W. J. Spencer	Xerox PARC
Secretary	Prof. K. S. Saraswat	Stanford U
Arrangements	Dr. N. M. Johnson	Xerox PARC
Proceedings	Dr. J. C. Knights	Xerox PARC
Program Chair	Dr. S. P. Keller	IBM Yorktown Heights
Program	Dr. R. A. Chapman	Texas Instruments
	Dr. B. A. Huberman	Xerox PARC, Stanford U
	Dr. R. W. Keyes	IBM Yorktown Heights
	Dr. J. Moll	Hewlett Packard
	Dr. V. Narayanamurti	Bell Labs
	Prof. W. G. Oldham	U.C. Berkeley
	Prof. J. D. Plummer	Stanford U
	Dr. G. E. Smith	Bell Labs
	Prof. E. D. Wolf	Cornell U

ICPS Committee Liaison

Secretary	Dr. R. Z. Bachrach	Xerox PARC, Stanford U
Treasurer	Dr. R. S. Bauer	Xerox PARC
Satellites	Dr. Sokrates Pantelides	IBM Yorktown Heights

We are grateful for the support of the ICPS Committee and to Irene Lile and Karen Martelli for their assistance in the running of the conference and in the preparation of the proceedings. We wish to particularly thank the Xerox Corporation for allowing us to use its facilities and services, and for acting as a gracious host during the conference.

September 26, 1984

J.C. Knights
Palo Alto, California

S.P. Keller
IBM
Yorktown Heights, New York

ACKNOWLEDGEMENTS

We would like to gratefully acknowledge the following organizations for the support which made this conference possible.

American Institute of Physics, American Physical Society, American Vacuum Society, IEEE, Optical Society of America, ATT Bell Labs, Eastman Kodak Company, Energy Conversion Devices, Exxon, Ford Motor Company Fund, General Electric, Gould, Inc., GTE, Hewlett-Packard Company, Hughes, IBM, Intel Corporation, ITT, Motorola, Inc., Philips Laboratories, RCA Labs, 3M Research, Varian, Xerox Corporation, Army Research Office, Air Force Office of Scientific Research, Defense Advance Research Agency, Department of Energy, Office of Naval Research, National Aeronautics and Space Administration, National Science Foundation, Solar Energy Research Institute.

Table of Contents

Preface v

Acknowledgements vii

PART I. OXIDES AND INSULATORS

CHARGE TRAPPING IN SiO_2
D. R. Young 1

GENERATION OF INTERFACE STATES BY INJECTION OF ELECTRONS INTO SiO_2
S. A. Lyon 8

THE DIFFUSION OF ION-IMPLANTED BORON IN SILICON DIOXIDE
J. Ng, J. F. Gibbons, and T. Sigmon 20

EFFECT OF HYDROGEN ON $Si-SiO_2$ STRUCTURES: AN ESR STUDY
W. E. Carlos and H. L. Hughes 34

PART II. SURFACES AND INTERFACES

EFFECT OF ANNEALING CHARGE INJECTION AND ELECTRON BEAM IRRADIATION ON THE $Si-SiO_2$ INTERFACE BARRIER HEIGHTS AND ON THE WORK FUNCTION DIFFERENCE IN MOS STRUCTURES
S. Krawczyk, M. Garrigues, and T. Mrabeut 39

DEPENDENCE OF Si–SiO$_2$ INTERFACE STRUCTURES ON CRYSTAL ORIENTATION AND OXIDATION CONDITIONS
T. Hattori, M. Muto, T. Suzuki, K. Yamabe, and H. Yamauchi 45

DYNAMIC CONDUCTIVITY MEASUREMENTS ON Si–SiO$_2$ INTERFACES DEGRADED BY OXIDE CHARGE AND AVALANCHE INJECTION
A. Zrenner and F. Koch 50

THE ATOMIC STRUCTURE OF Si–SiO$_2$ INTERFACES SUGGESTING A LEDGE MECHANISM OF SILICON OXIDATION
J. H. Mazur and J. Washburn 52

Part III. INTERCONNECTIONS AND SILICIDES

LASERS FOR INTERCONNECTIONS, CIRCUIT FABRICATION AND REPAIR
R. J. von Gutfeld 56

A STUDY OF LOW TEMPERATURE Pd AND Ni SILICIDES FORMED ON DRY ETCHED SILICON SURFACES
X. C. Mu, A. Climent, and S. J. Fonash 63

Part IV. LITHOGRAPHY

LITHOGRAPHY REQUIREMENTS IN COMPLEX VLSI DEVICE FABRICATION
A. D. Wilson 69

CORRECTION OF ELECTRON BEAM PATTERNS FOR EXPOSURE AND PROCESS EFFECTS
P. Hendy, M. E. Jones, P. G. Flavin, and C. Dix 92

Part V. REACTIVE ION ETCHING

PLASMA-ASSISTED ETCHING: ION-ASSISTED SURFACE CHEMISTRY
J. W. Coburn and H. F. Winters 98

DAMAGE EFFECTS IN REACTIVE ION ETCHING
S. J. Fonash 106

SIMULATION AND ANALYSIS OF ANOMALOUS PLASMA ETCHING
S. Kawazu, T. Nishioka, and H. Koyama 120

Part VI. DEFECTS IN SEMICONDUCTORS

DEFECTS IN SILICON AND THEIR RELATION TO VLSI
S. T. Pantelides 125

MICRODEFECT INTRODUCTION TO ENHANCE VLSI WAFER PROCESSING
G. B. Larrabee 139

DEFECT CONTROL AND UTILIZATION IN SILICON SUBSTRATES FOR VLSI
Xu, K., Wang, W., Shi, Z., Luo, G., He, D., Shao, H. 145

Part VII. DEVICES

THE PHYSICS OF SCALED BIPOLAR DEVICES
T. H. Ning 151

FUNDAMENTAL LIMITATIONS ON DRAM STORAGE CAPACITORS
W. P. Noble and W. W. Walker 156

MONTE CARLO SIMULATION OF CONTACTS IN SUBMICRON DEVICES
P. Lugli, U. Ravaioli, and D. K. Ferry 162

THE INFLUENCE OF HOT CHANNEL ELECTRONS ON THE SURFACE POTENTIAL IN MOSFETs
D. Schmitt-Landsiedel and G. Dorda 167

PHYSICS AT THE LIMITS OF VLSI SCALING
P. M. Solomon 172

THERMALLY INDUCED TRANSITION METAL CONTAMINATION OF SILICIDE SCHOTTKY BARRIERS ON SILICON
A. Prabhakar and T. C. McGill 181

MOBILITY LIMITATIONS IN VLSI DEVICES DUE TO THE FIELD-BROADENING EFFECT
V. K. Arora 186

SCALING PROBLEMS AND QUANTUM LIMITS IN INTEGRATED CIRCUIT MINIATURIZATION
H.-U. Habermeier 192

Part VIII. MATERIALS PROCESSING

THE APPLICATION OF SILICON MOLECULAR BEAM EPITAXY TO VLSI
J. C. Bean 198

PROCESS-INDUCED MICRODEFECTS IN VLSI SILICON WAFERS
F. Shimura and R. A. Craven 205

TEMPORAL STRUCTURE OF Si (111) SURFACE OXIDATION
K. S. Yi, W. Porod, R. O. Grondin, and D. K. Ferry 220

GETTERING FOR VLSI
G. F. Cerofolini and M. L. Polignano................................225

Part IX. PROCESS MODELLING

TWO-DIMENSIONAL NUMERICAL OF THE CHARGE TRANSFER IN
GaAs CHARGE-COUPLED DEVICES OPERATED IN THE GHz RANGE
D. Sodini, A. Touboul, and D. Rigaud 240

MODELLING THE INITIAL REGIME OF DRY THERMAL OXIDATION OF
SILICON
G. Ghibaudo and A. Fargeix..247

Part X. DESIGN AUTOMATION, DEVICE LAYOUT AND PACKAGING

PHYSICAL INTUITION AND VLSI PHYSICAL DESIGN
R. Linsker 251

CONCEPTS OF SCALE IN SIMULATED ANNEALING
S. R. White..261

Part XI. III-V COMPOUNDS

HIGH SPEED GALLIUM ARSENIDE TRANSISTORS FOR LOGIC APPLICATIONS
L. F. Eastman 271

ELECTRON SCATTERING AND MOBILITY IN A QUANTUM WELL HETEROLAYER
V. K. Arora and A. Naeem 280

CHARACTERIZATION OF DEEP LEVEL DEFECTS IN REACTIVE ION BEAM ETCHED InP
Y. Yuba, K. Gamo, Y. Judai, and S. Namba 286

Part I. Oxides and Insulators

Charge Trapping in SiO$_2$

Donald R. Young
IBM Thomas J. Watson Research Center
Box 218, Yorktown Heights, New York 10598

ABSTRACT

The importance of electron and hole trapping in SiO$_2$ have been well documented as important considerations in the design of MOS-Field Effect Transistors. These considerations become increasingly important as the channel width is decreased. In addition, the techniques used to make contemporary-small devices require the use of radiation which generates new trapping sites in the SiO$_2$. The traps are classified in terms of those located in the bulk of the SiO$_2$ and those located at the Si-SiO$_2$ interface. The process dependence of the bulk traps is entirely different than the process dependence of the interface traps. It has been observed that prolonged heat treatments in N$_2$ at 1000° C decrease the bulk trap density but increase the interface trap density and that this increase in interface trap density can be reduced by the use of a O$_2$ anneal which is also done at 1000° C. A recent observation by Aslam, Balk and Young shows a relationship between the deep interface trap with a cross section of 5×10^{-19}cm^2 and the shallow trap observed only at low temperatures with a cross section of 1×10^{-16}cm^2. This suggests a bimodal model to describe these traps as arising from a single site.

INTRODUCTION

The decrease in the size of MOSFET's, to obtain higher packing densities, leads to shorter channel widths. This decrease in channel width increases the population of electrons in the range of 3eV which can surmount the Si-SiO$_2$ barrier, flow into the SiO$_2$ and have a possibility of being trapped, with a resultant electric field that is applied to the device. This field counteracts the applied field and degrades the device characteristics. As the SiO$_2$ thickness is reduced, the electron traps associated with the Si-SiO$_2$ interface become increasingly important as compared with the traps located in the bulk of the SiO$_2$. The techniques used to make small devices apply radiation to the SiO$_2$ and this radiation can generate new traps in the SiO$_2$[1]. The structures used for small devices are becoming increasingly complex and frequently involve interfaces between insulators that can trap impurities that can have a large effect on the electron trapping rate[2].

---

[1] L.M. Eprath, D.J. DiMaria, Solid State Technology, April 1981.
[2] D.J. DiMaria, W. Reuter, D.R. Young, F.L. Pesavento, and J.A. Calise, J. Appl. Phys. 52, 1366 (1981).

TABLE I

	Cross section (cm²)	Trap description
Bulk traps		
	1×10^{-17}	Electron trap water related[3]
	2×10^{-18}	Electron Trap Water related[4]
Interface Traps		
	$10^{-15} - 10^{-16}$	shallow electron trap [5,6]
	5×10^{-19}	electron traps [5]
	$10^{-13} - 10^{-14}$	hole traps[7]

Table I. Classification of Traps as Bulk Traps and Interface Traps

In order to screen out the effects of asperities on thin oxide devices it may be necessary to deliberately add traps such as the use of silicon nitride layers or oxinitrides[8,9]. In addition, the processes used are becoming complex and offer many more possibilities for introducing trapping sites into the SiO_2. As a result, it is evident that the trapping characteristics of SiO_2 continue to be important technologically. On a more scientific basis the understanding of the microscopic nature of the trapping sites is still in a primitive state and further work is required.

II. CLASSIFICATION OF TRAPS

It is helpful to classify the traps in terms of the traps associated with the bulk of the SiO_2 and those associated with the Si-SiO_2 interface. The difference in the response to high temperature processing conditions of the bulk and the interface traps taken from earlier work is given in fig. 1.[10]

The electron trapping at 295°k is dominated by bulk traps and the trapping at 77°k is dominated by shallow interface traps. For this experiment an electron current is induced through

---

[3] E.H. Nicollian, C.N. Berglund, P.F. Schmidt, and J.M. Andrews, J. Appl. Phys. 42, 5654 (1971).
[4] A. Ushirokawa, E. Suzuki, and M. Warashima, Jap. J. Appl. Phys. 12, 398 (1973).
[5] D.R. Young, J. Appl. Phys. 56, 4090 (1981).
[6] T.H. Ning, J. Appl. Phys. 49, 5997, (1978).
[7] J.M. Aitkin and D.R. Young, IEEE Transact. Nuclear Sci. NS-24, 2128 (1977).
[8] D.J. Dimaria, D.R. Young, and D.W. Ormond, Appl. Phys. Lett., 31, 680 (1977).
[9] T.W. Ekstedt, S.S. Wong, Y.E. Strausser, J. Amano, S-H. Kwan, and H.R. Grinolds, Insulating Films on Semiconductors, published by North-Holland, edited by J.F. Verwey and D.R. Wolters (1983).
[10] D.R. Young, E.A. Irene, D.J. DiMaria, R.F. Dekeesmaecker H.Z. Massoud, J. Appl. Phys. 50, 6366 (1979).

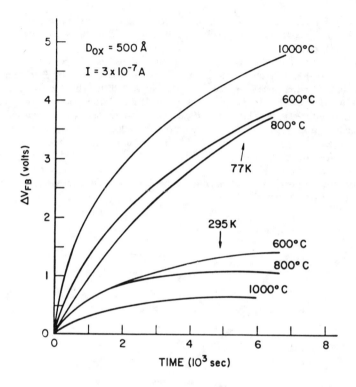

Figure 1. Electron Trapping at 295°K and 77°K as a function of time.

the SiO$_2$ by the avalanche injection technique and the electron trapping is measured by measuring the charge build up in the SiO$_2$ using using the high frequency C-V shift technique [11].

---

[11] For a discussion of these techniques see MOS Physics and Technology, E.H Nicollian, and J. R. Brews, Published by Wiley-Interscience, 1982.

These results show a decrease in electron trapping resulting from the annealing treatment at 1000°C but the interface trapping is minimized by a treatment at 800°C.

III. BULK ELECTRON TRAPS

The original work of Nicollian et al. [3] showed that the bulk traps are due to water related species in the SiO_2 Their conclusions were subsequently supported by the work Gdula [12] and by more recent work of Feigl et al. [13]. It has been shown that the trap cross section for a sample after aluminum metallization is $1 \times 10^{-17} cm^2$ but decreases to $2 \times 10^{-18} cm^2$ if the sample is heated is to 400°C after metallization. The subsequent work by Hartstein and Young [14] using multiple reflectance IR spectroscopy has suggested that the $1 \times 10^{-17} cm^2$ trap is due to SiOH and that the $2 \times 10^{-18} cm^2$ trap is due to loosely bound H_2O. The electron trapping rate can be reduced to very low values if careful procedures are used to grow and heat treat the SiO_2 under ultra dry conditions, long treatments at 1000°C in N_2 are used and if the post metallization treatment at 400°C is also used [15]. Recent unpublished work by Weinberg, Young and Cohen has shown that surprisingly short times of 10 sec at 600°C can "dry out" a sample that had been previously exposed to a H_2O ambient. These samples after this treatment are significantly improved, but not as good as the samples made with the long time N_2 anneal. These results are given in fig. 2.

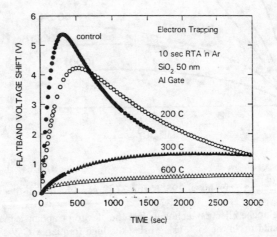

Figure 2.   Electron Trapping as a function of time.

---

[12]   R.A. Gdula, J. Electrochem. Soc., 123, 42 (1976).
[13]   F.J. Feigl, D.R. Young, D.R. DiMaria, S. Lai and J. Calise, J. Appl. Phys. 52, 5665 (1981)
[14]   A. Hartstein and D.R. Young, Appl. Phys. Lett., 38, 631 (1981).
[15]   S.K. Lai, D.R. Young, J.A. Calise, and F.J. Feigl, J. Appl. Phys. 52, 5691 (1981).

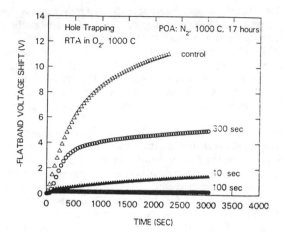

Figure 3. Hole Trapping as a function of time for various annealing treatments at 1000°C.

IV. INTERFACE TRAPS

The initial results on the shallow electron traps described previously[5] suggest that their response to high temperature annealing treatment is similar to that of hole traps which control the radiation hardness. Subsequent work has confirmed this conclusion and in addition it has also been observed that the $5 \times 10^{-19} cm^2$ electron trap also reacts in the same way. This work was performed by Aslam, Balk and Young, [16]. There had been a model proposed [17] indicating that the hole traps might be due to trivalent silicon in the SiO_2. It is reasonable that this concentration of the trivalent silicon sites might increase if the sample were heated at high temperature for a long time in a non oxidizing ambient. This has been observed.

In addition, it can be expected that a short oxygen treatment might reverse this effect. This was also observed by Aslam et al. and confirmed by recent unpublished work by Weinberg, Young and Cohen is shown in Fig. 3. The long post annealing treatment (17 hours at 1000°C in N_2 was used to initially increase the hole trapping rate. This occurred; however, an expected result was that the combination of this treatment in N_2 and the short O_2 anneal gave the best overall result, if the O_2 anneal was only applied for a short time of 100 sec. It is

---

[16] M. Aslam, P. Balk, and D.R. Young to be published J. Elect. Devices.
[17] C.M. Swensson, in The Physics of SiO_2 and its Interfaces, S.T. Pantelides, Ed., Pergamon, New York, 1978.

suggested that the oxygen anneal has the desirable effect of removing trivalent silicon or excess silicon defects in the SiO_2 but also creates new defects as additional oxide is grown. It is well known that the presence of nitrides (observed by Aslam et al and Weinberg et al.) at the Si-SiO_2 interface inhibits further oxide growth[18,19]. This model explains the improvement resulting from the combination of the N_2 treatment and the short O_2 treatment and also indicates why the optimum result occurs for a short treatment. The net result in the limit of long O_2 treatment would be the same as for a sample withdrawn from the furnace in the O_2 ambient. These results for hole trapping are also observed for the interface electron traps.

An additional observation based on the work of Aslam et al[16] demonstrated that there exists a close connection between the shallow electron traps that are only observed at temperatures below room temperature and the $5 \times 10^{-19} cm^2$ traps that can be filled at room temperature. This conclusion was based on the following experiment. The shallow electron traps were partially filled at 77°K which could be readily accomplished with a relatively small electron fluence due to the large cross section for these traps. The sample was then heated to room temperature and a much larger fluence applied to enable us to investigate the $5 \times 10^{-19} cm^2$ trap. It was observed that about ½ of these traps had been previously filled by the preceding 77°K treatment. These results suggest that at 77°K the electrons are first captured by the shallow levels and then some of them tunnel into the deeper levels. This suggests a bimodal trap as shown schematically in fig. 4. This figure clearly indicates the reason for the relatively large cross section observed at 77°K ($10^{-16} cm^2$) and the relatively small cross section (5×10^{-19}) observed at room temperature.

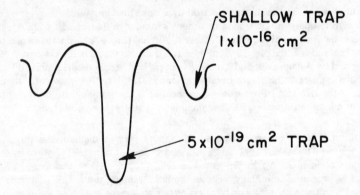

Figure 4. Schematic representation of bimodal trap shown potential as a function of distance with a shallow minimum and a deeper minimum.

---

[18] S.I. Raider, R.A. Gdula, and J.R. Petrak, Appl. Phys. Lett. 27, 150 (1975).
[19] W.J.M.J. Josquin and Y. Tawrminga, J. Electrochem. Soc. 129, 1803 (1982).

V. CONCLUSIONS

The electron traps located in the bulk of the SiO_2 characterized by cross section of 1×10^{-17} and 2×10^{-18} are water related sites due to SiOH and loosely bound H_2O respectively. Their density can be reduced by prolonged high temperature (i.e. 1000°C) treatment in N_2

The Si-SiO_2 interface traps result from a O_2 deficiency, and their density can be reduced by short 1000°C anneals in O_2 The electron traps are characterized by cross sections of $10^{-15} - 1 - ^{-16} cm^2$ and $5 \times 10^{-19} cm^2$. The $10^{-15} - 10^{-16} cm^2$ traps are shallow traps and can only be filled at temperatures below room temperature. The shallow traps and the 5×10^{-19} traps are closely related and a bimodal characteristic has been proposed. In addition, hole traps with cross sections of $10^{-13} - 10^{-14} cm^2$ are observed that also respond in a similar way to high temperatures annealing treatments.

VI. ACKNOWLEDGEMENTS

The author is indebted the following colleagues, J. Calise, D. J. DiMaria, F. Feigl, S. Lai, R. DeKeersmaecker J. Aitken, E. Irene, M. Aslam, P. Balk.

GENERATION OF INTERFACE STATES

BY INJECTION OF ELECTRONS INTO SiO_2

by

S.A. Lyon
Department of Electrical Engineering and Computer Science
Princeton University
Princeton, NJ 08544

ABSTRACT

Several techniques have been used to inject electrons into SiO_2 with various energies. Interface states are found to be generated whenever electrons flow through the oxide. However, the efficiency of interface state generation depends upon the method of electron injection. At high enough fields, positive charge is produced in the oxide which enhances the production of interface states. All of the states are amphoteric and are probably dangling Si bonds at the interface (P_b-centers).

1. INTRODUCTION

It is probably not an overstatement to say that the success of silicon technology is due in large measure to the dielectric properties of silicon dioxide. In particular, the interface between Si and SiO_2 is critical for modern devices. As lateral dimensions are reduced, devices must also shrink in the vertical direction. The electrical quality of this interface is important, even for nominally bulk devices because of their small vertical extent. In bipolar transistors interface states provide generation-recombination centers which reduce the current gain, generate noise, and increase collector leakage currents [1]. In a MOSFET (Metal-Oxide-Semiconductor Field-Effect Transistor) interface states will shift the threshold voltage, reduce the transconductance both by trapping electrons in the channel and by reducing their mobility through scattering [2], and generate 1/f noise [3].

It has been known since the mid 1960's that the oxidation of silicon in dry O_2 produces a large density of interface states ($\sim 10^{12} cm^{-2} eV^{-1}$), but that the addition of water vapor to the oxidizing ambient [4,5] or annealing in hydrogen [6] will reduce the interface state density to acceptable levels ($\sim 10^{10} cm^{-2} eV^{-1}$). Until recently, there has been little concern about the generation of interface states during device operation except for specialized applications such as high radiation environments. It has been known for some time that ionizing radiation causes interface states to appear [7,8].

Recently, it has become clear that a better understanding of the Si-SiO_2 interface is necessary. As devices are made smaller, electric fields within them increase. It has been found that electrons in the channel of MOSFETs can acquire sufficient energy (~ 3.1 eV) to be injected into the SiO_2, creating interface states [9]. Similarly, electrons are accelerated by the large collector-base field in bipolar transistors and can be injected into the SiO_2. In addition, nonvolatile memory devices store charge on a floating gate or purposely introduced electron traps. In order to write and erase these devices, electrons must pass through the oxide. As a result of unwanted electron trapping and interface state generation, the number of write-erase cycles is limited. Thus it can be seen that degradation associated with hot electron injection into SiO_2 is one of the most important limiting factors for highly integrated circuits [10].

In this paper we will discuss recent results on the generation of interface states when electrons pass through the oxide. The problem of electron trapping is reviewed in this conference by Young [11]. In Sec. 2 we briefly review some of the properties of interface states. In Sec. 3 we will discuss the interface states produced by various electron injection techniques, including Fowler-Nordheim tunneling (high-field stress), internal photoemission, and avalanche injection. The results will be summarized in Sec. 4.

2. Si–SiO$_2$ INTERFACE STATES

Following the usual convention, when we discuss interface states we mean states positioned energetically within the Si bandgap which can exchange charge with the silicon. We will only be concerned with "fast" states. Avalanche injection [12,13] and Fowler-Nordheim tunneling also produce "slow" states which exchange charge with the silicon on a time scale from minutes to days. These states should not be confused with slow states associated with sodium or other mobile ion contamination. Modern processing techniques have reduced the Na$^+$ level to the point that it is usually not important.

One of the goals of interface studies has been the development of detailed microstructural models of the defects responsible for the states. Recent Electron-Spin-Resonance (ESR) experiments [14,15] have identified a center, first observed by Nishi [16,17] as "trivalent silicon" at the Si-SiO$_2$ interface. On a (111) surface the "trivalent silicon" center, labeled P_b, is an Si atom bonded to three other Si atoms, with a dangling bond pointing into the SiO$_2$. Similar structures exist for other surface orientations. Observation of the hyperfine structure of the P_b center has confirmed this structural model [18]. In an extensive study, Poindexter et al. found a strong correlation between the density of P_b centers and the interface-state density of "as grown" oxides on both (111) and (100) surfaces [15].

Lenahan and co-workers have also shown that P_b centers are created when MOS devices are exposed to ionizing radiation [19,20]. Again, a good correlation is obtained between the rate of generation of interface states and P_b centers [19], as well as their annealing characteristics [20].

Recent experiments combining electrical and ESR measurements have established that the P_b-center is an interface state [20-24]. From variations of the ESR signal with gate bias [20-23], it can be shown that the P_b-center does exchange charge with the silicon. These studies show that the P_b center can capture either one or two electrons, making it an amphoteric center. In addition, Spin-Dependent-Recombination measurements [24] have demonstrated that this trivalent silicon can act as a generation-recombination center.

While the ESR results definitely show that the P_b-center is an interface state, they cannot answer the question of whether that is the only defect responsible for interface states in a "good" sample. (Sodium ions in the SiO$_2$ are known to act as interface states). Due to uncertainties in the ESR measurement, it is difficult to make correlations between electrically measured interface-state densities and P_b-center densities more quantitative than about a factor of 2. This leaves open the possibility that some impurity or other defect could be an interface state. Modern silicon processing incorporates significant quantities of hydrogen and chlorine in the SiO$_2$ and numerous models, including strained [25,26] or weak bonds [27] at the interface, charged centers in the oxide [28], and Anderson localization of carriers caused by random potential variations at the interface [29,30] have been proposed to explain the interface states. Recent measurements, which will be discussed below, show that for certain experiments all of the interface states in the middle portion of the Si bandgap ($\gtrsim 0.2$ eV from either band edge) are almost certainly P_b-centers.

3. ELECTRON INDUCED STATES

For simplicity, MOS capacitors were used in the studies we will describe. Both n- and p-type substrates were used. Heavily doped p-type substrates were necessary for the avalanche injection experiments. The oxides were generally grown by a "standard" process on (100) surfaces to a thickness of 500 - 2000Å (oxidation in dry O_2 with a few percent HCl at 1000°C, a 30 min. anneal in N_2 at 1000°C, evaporation of Al for back contact and 135Å semi-transparent field plate, and a final hydrogen or forming gas anneal at ~450°C).

Many of the experiments we will discuss below were performed at or near liquid nitrogen temperature. Often this approach can help simplify the interpretation since processes like atomic diffusion will be frozen out. For example, it is known that interface states do not form after exposure to ionizing radiation if the sample is kept at low temperature [32-35]. In order to measure the interface state density at low temperature, a method was developed by Jenq [34,36] which is a modification of one described by Gray [37]. This technique uses low-temperature Capacitance-Voltage (C-V) curves as illustrated in Fig. 1. Taking the dotted curve, the trace is started at positive gate voltage which accumulates majority carriers (electrons in this case) at the interface. The applied voltage is ramped negative and the high-frequency (1 MHz) capacitance drops at about 3V, as electrons are pushed away from the interface, depleting the Si. However, at the low temperature, electrons are unable to escape from interface states that are deeper than about 0.2 eV below the Si conduction band edge. This means that the interface is in its most negative condition. The downsweep is the lower curve, and the capacitance becomes very small for no inversion layer forms at 90 K, forcing the Si into deep depletion. The sweep was stopped at -5 V, and the sample illuminated briefly to generate minority carriers (holes) forming an inversion layer. This causes the capacitance to rise as shown. Now the gate voltage is ramped back toward accumulation, giving the upper dotted curve. A ledge is apparent at about -3 V. Here free holes have been swept from the interface, but now holes are frozen into the interface states leaving them in their most positive condition. The capacitance rises as the depletion region narrows, but then stays relatively constant as electrons from the bulk reach the interface and annihilate the holes in the interface states. As the gate voltage is swept more positive, all the holes are neutralized, and the two curves join together in accumulation.

Since in the downsweep the interface states are filled with electrons, while on the upsweep they were filled with holes, we can determine the total charge stored at the interface by the width of the "ledge." Measuring the voltage difference (ΔV) between the step at the left side of the ledge (-2.8 V) and at the *same capacitance* on the downsweep (+ 1.2 V) we can determine the charge stored at the interface

$$Q_{it} = C_{ox}\Delta V \tag{1}$$

where C_{ox} is the oxide capacitance. This can be put in terms of the energy-integrated (over middle 0.7 eV of the Si bandgap) interface state density

$$N_{it} = Q_{it}/q \tag{2}$$

where q is the electronic charge.

The solid curve in Fig. 1 was taken on a fresh sample. The narrow ledge shows that the initial interface state density is very small. The dashed curve in

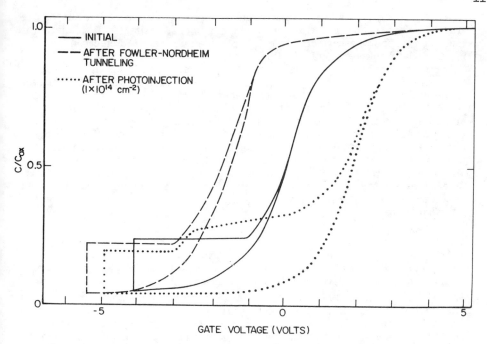

Figure 1: Low Temperature C-V curves showing the effect of high-field stress followed by electron injection. The sample had an n substrate and a dry HCl oxide. Solid curve: fresh sample. Dashed curve: after Fowler-Nordheim tunneling 3×10^{-3} C/cm^2 from the substrate. Dotted curve: after photoinjecting 1×10^{14} electrons/cm^2. From [31].

Fig. 1 was taken after the capacitor was subjected to a "high-field stress" at 90 K. A field of ~ 7 MV/cm was applied across the oxide (gate positive) which causes Fowler-Nordheim tunneling of electrons out of the Si into the conduction band of the SiO$_2$ [38]. The current density was approximately 1.7×10^{-6} A/cm^2, and about 3×10^{-3} C/cm^2 was passed through the oxide [31]. We can see that the C-V curve has shifted to the left, indicating that positive charge has been trapped in the SiO$_2$, and it is known to reside within about 30Å of the Si-SiO$_2$ interface [39,40]. The dotted curve in Fig. 1 does not have a clearly defined ledge, indicating that very few interface states have formed. This is similar to the situation encountered with radiation damage, in that positive charge is trapped at low temperature, but interface states do not form until the sample is warmed up [33,35,36,40,41].

In this experiment [39] the sample was held at 90 K and electrons were photoinjected from the Al field plate into the SiO$_2$. After injecting $\sim 10^{14}$ electrons/cm^2, changes in the C-V curve saturated at the dotted curve of Fig. 1. From the width of the ledge it is apparent that many interface states have been generated. More importantly we see that the left side of the ledge coincides with the curve (dashed) obtained immediately after the high-field stress. This means that with positive charge frozen into the interface states, the charge there is the same as before the interface states were generated. In addition, the ledge is symmetrical about the initial (solid) curve. We believe that photoinjecting electrons until saturation has annihilated all of the positive charge, and thus this symmetry of the ledge implies that the interface states can trap as many electrons as holes, i.e. there are equal numbers of donor and acceptor-like states. This is almost certainly not a coincidence. The simplest explanation is that each trapped positive charge can capture a photoinjected electron, producing a single amphoteric defect. It is known from a preliminary ESR study [42] that Fowler-Nordheim tunneling produces P_b-centers. Moreover, the correlation between radiation induced interface states and P_b centers is well established, as is the equivalence between the states generated by high-field stress and radiation [33,36,40,43,44]. Coupled with the fact that the P_b-center is known to be an amphoteric interface state, we conclude that each trapped positive charge becomes a P_b center when it captures an electron at low temperature. A possibly similar production of interface states by electron capture has been observed at room temperature after the injection of holes into SiO_2 by an avalanche [45]. The high-field stress experiment shows that essentially all (> 90%) of the interface states generated in this manner are interfacial Si dangling bonds. It is not necessary to invoke any defect other than the P_b-center to account for the interface states.

Another technique for injecting electrons into SiO_2 is internal photoemission [46]. This method has the advantages that electrons can be injected from either interface, and that the average electric field in the oxide can be varied independently of the current. The UV light needed to excite electrons for photoinjection appears to have no effect in the absence of a current [47].

It has been found that interface states are generated when large numbers of electrons are passed through the SiO_2 using photoinjection [47,48]. This generation is shown in Fig. 2, where interface-state density is plotted against the electron fluence. It should be noted that many more electrons pass through the oxide in this experiment than in the Fowler-Nordheim tunneling studies. Unlike the experiments discussed above, interface states are generated immediately at low temperature during photoinjection. In fact, for "standard" oxides the generation rate (states/coulomb) is essentially independent of temperature [47] (90 - 300 K) and direction of current flow [48]. There is only a small change in interface-state density when samples are warmed from 90 K to 300 K. It has been shown that oxides grown in steam or ones that did not receive a low temperature hydrogen anneal generate interface states much more rapidly than "standard" oxides when injected at room temperature [49], though there is less than a factor of 2 difference for low temperature injection [47].

From Fig. 2 it can be seen that interface states are produced more rapidly in a thicker oxide. The generation rate increases approximately linearly with oxide thickness from 500 - 1800Å [47]. The origin of this thickness dependence is not clear. It is highly unlikely that some species diffuses from the bulk to the interface because of the low temperature and the fact that it would have to be neutral to account for the independence of the generation rate on the sign of the applied field. It has been suggested that interfacial stress may account for the thickness dependence [47]. Stress has also been implicated in radiation damage produced by high-energy electrons [50].

A log-log plot of interface-state generation during photoinjection at low temperature is shown in Fig. 3. For the same sample preparation, the interface-state density increases with average oxide field between 1 and 2 MV/cm. It is also clear that the generation does not saturate. Rather, it appears that the interface-state density increases as about one-half to two-thirds power of the injected fluence.

The marked differences between the high-field stress and photoinjection experiments could be interpreted as the formation of different kinds of defects. However, recent room-temperature photoinjection experiments by Mikawa and

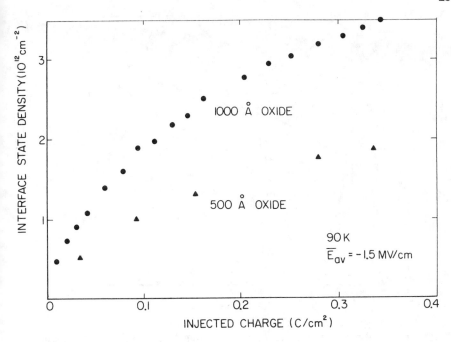

Figure 2: Number of interface states generated as a function of the charge passed through the oxide at 90 K. Average bias field was -1.5 MV/cm. The symbols Δ and ■ represent p-type MOS capacitors with 500- and 1000-Å dry HCl oxides, respectively. From [47].

Lenahan [51] show that P_b centers are created, though there does not appear to be a build-up of positive charge. Thus, the passage of electrons through the SiO_2 produces P_b centers in a different manner than ionizing radiation or Fowler-Nordheim tunneling.

Another technique for flowing an electron current through SiO_2 is to use avalanche injection. This method developed as a convenient way of injecting electrons for the purpose of studying trapping in the SiO_2 [52,53]. It does not have the problems of optical detrapping seen with photoinjection or high-field detrapping in the Fowler-Nordheim experiments. The MOS capacitor is made on a heavily doped Si wafer and a high voltage RF (radio frequency, typically 0.1 - 1 MHz) is applied across the oxide. On the positive swing (for a p-type substrate) of the gate voltage, the Si is driven into deep depletion. For a large enough RF voltage, the silicon avalanches, producing hot electrons which are injected into the SiO_2 (using an n-type substrate, holes can be injected). On the reverse swing of the RF bias the surface is accumulated, destroying the inversion layer produced by the avalanche so that the capacitor is ready for the next injection cycle. As electrons are trapped in the SiO_2, the current will change. A constant current can be maintained with a feedback circuit to adjust a DC bias [52] or the RF voltage [12]. High average currents are possible with this injection method, and fluences up to tens of Coulombs/cm^2 can be obtained in a reasonable time.

Interface-state formation was observed in some of the early studies [53]. More recently we have investigated their generation both at room temperature and low temperature [54]. The solid curve in Fig. 4 shows how the interface-state density (measured using the low-temperature technique) increases with

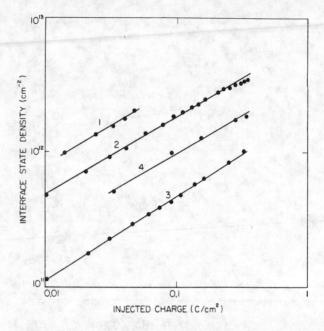

Figure 3: Number of interface states generated as a function of injected charge plotted on log-log scale. All the samples shown had p-type substrates and were photoinjected at 90 K. Curve 1: 1000-Å dry HCl oxide, biased with -2 MV/cm. Curve 2: 1000-Å dry HCl oxide, biased with -1.5 MV/cm. Curve 3: 1000-Å dry HCl oxide biased with -1 MV/cm. Curve 4: 950-Å wet oxide, biased with -1 MV/cm. The slopes using linear least-squares fit are 0.56, 0.57, 0.64 and 0.61, respectively. From [47].

electron fluence. The lower points show the increase in interface states while injecting at low temperature. Injection of sample A was terminated after 2.4 C/cm^2 had passed through the oxide. The sample was warmed to room temperature for 40 hr and recooled to measure interface-state density, with the result marked A. Samples B and C were treated in a similar manner, but injection was stopped after different charge fluences. Each sample developed additional states as shown by the points marked A, B, and C. At low temperature the increase in interface states follows a power law similar to that seen for photoinjection, though the generation rate is lower for the avalanche injection.

The avalanche injection experiments can be interpreted in terms of interface-state formation through two processes. The passage of electrons through the interface produces states immediately at low temperature via the same mechanism as in the photoinjection experiments. At the same time, some of the hot electrons in the avalanche produce positive charge in the SiO_2, quite possibly by injecting holes at the SiO_2-Al interface [54]. The positive charge can then form interface states on warming as observed for high-field stress and radiation damage. It has also been observed that photoinjecting a small number of electrons after avalanche injection and while the sample is still cold produces more interface states. This behavior is similar to that discussed above in the high-field stress experiments. A build-up of positive charge is known to occur during room-temperature avalanche injection [12,13] though at low temperature

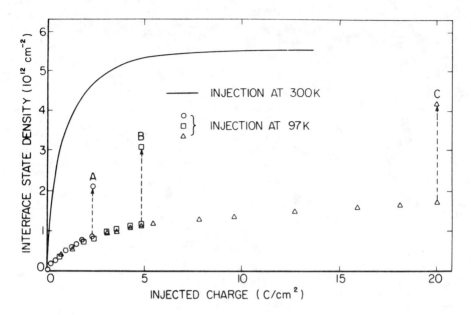

Figure 4: Interface-state density vs. injected charge for avalanche injection at both 300 and 97 K. Three samples A(○), B(■), and C(△) had electrons avalanche injected into the oxide at 97 K, and periodically the energy-integrated interface-state density was measured by the Jenq method. The avalanche injection was stopped after 2.4, 4.8 and 20 C/cm^2, respectively. Each sample was then warmed to room temperature for 40 hrs. and the interface-state density was found to have increased as indicated by the dotted arrows. The solid curve shows how the interface-state density (as measured by the Jenq technique) increased for a sample which had been avalanche injected at room temperature. From [54].

it may be masked by electron trapping in shallow traps [12].

If the above interpretation of the avalanche injection studies is correct, it should be possible to observe both interface-state generation mechanisms by photoinjecting electrons with a large oxide field. The results of this experiment are shown in Fig. 5. Electrons were photoinjected from the Al field plate at 90 K with oxide fields between 6 and 7.2 MV/cm. The current density was about 2.4 × 10^{-6} A/cm^2 and the total photoinjected charge was 4.3 × 10^{-3} C/cm^2. At the highest fields there is a component of the current from Fowler-Nordheim tunneling, but it was never more than 10% of the total current. The flat-band voltage shift is plotted in the upper panel of Fig. 5. For the two lower fields, the flat-band shift is positive, indicating negative charge trapping in the oxide. At the highest field, however, the flat-band shift becomes negative, which shows that positive charge is being trapped. The lower panel of Fig. 5 shows the effect of field on the interface-state generation. As the average electric field is increased, the efficiency of producing interface states at low temperature is reduced. The number of new states that form when the sample is warmed to 290 K for 8 hrs. increases with field, however. This result is expected, given the fact that more positive charge is trapped at higher fields.

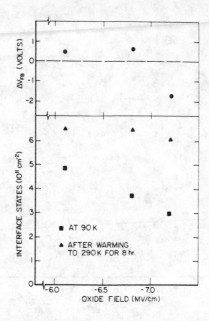

Figure 5: Flat-band shift and interface state density for p-type MOS capacitors with a 1000Å dry HCl oxide as a function of electric field during photoinjection at 90 K. Electrons were injected from the field plate to a fluence of 4.3×10^{-3} C/cm^2.

The distribution of interface states produced during photoinjection at various average fields was measured using the quasi-static method [55] and is shown in Fig. 6. The photoinjected current density was 1.2×10^{-6} A/cm^2 and the total charge fluence was 8×10^{-4} C/cm^2. The lowest field produced only a rising background in the upper half of the Si bandgap. (The steep increases at both sides of the figure are artifacts of the measurement.) By 6.6 MV/cm, a peak in the interface-state density is beginning to appear at about 0.7 eV above the valence band edge. At 7.0 MV/cm there is a large peak at 0.7 eV above the valence band edge. The curve had to be reduced by a factor of 3 in order to fit on the figure. A peak in the interface state density at this position is often observed after radiation damage [50], high field stress [40], and avalanche injection [56]. This peak appears to be characteristic of processes which introduce positive charge into the SiO$_2$ that subsequently causes the formation of interface states.

4. CONCLUSION

The injection of hot electrons into SiO$_2$ may place a severe limitation on the scaling of silicon devices until a way is found for preventing the formation of interface states. From a number of experiments it has been learned that interface states are produced whenever electrons pass through the interface. All these states appear to be the same defect, an Si dangling bond at the interface. However, these defects can be generated in at least two ways. One can occur at relatively low fields and only requires electrons to pass through the interface. The second is more efficient (per electron) at generating interface states, but requires high enough fields and hot enough electrons to produce positive charge

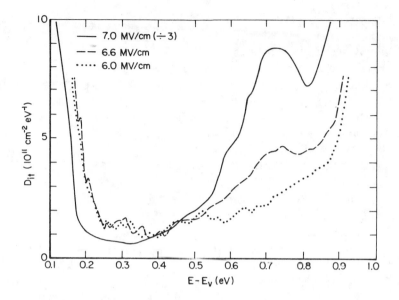

Figure 6: Distribution of interface states produced by photoinjecting electrons from the field plate of an MOS capacitor. The oxide was grown to 1000Å in dry O_2 and HCl on a p substrate. The electrons were injected at 90 K to a fluence of 8×10^{-4} C/cm^2 with various applied fields. The solid curve is reduced by a factor of 3.

in the oxide. The details of these mechanisms are not yet understood.

One might think that electron injection into SiO_2 could be eliminated by reducing all voltages below about 3 volts since the barrier to electrons crossing the interface from the Si to the SiO_2 is 3.1 eV. However, this assumes that electron-electron scattering in the channel of a MOSFET would be unable to produce significant numbers of electrons with kinetic energies larger than the potential drop from source to drain. In addition, recent Monte Carlo simulations indicate that the effective barrier is reduced by collisional broadening [57]. The calculations appear to show that the voltages would have to be reduced well below 2.5 V to eliminate hot electron injection. In summary, there may not be an "easy fix." It will probably be necessary to understand the interface-state generation mechanisms in order to get away from the degradation caused by hot electrons.

ACKNOWLEDGEMENTS

This work has grown out of the efforts of many colleagues, especially Walter C. Johnson who introduced me to this field, as well as G. Hu, S. Pang, J.K. Wu, T. Sunaga, D. Sola, and S.T. Chang, to whom I would like to express my gratitude. I would also like to thank P.M. Lenahan for a preprint of Ref. 51, and C. Agans for her expert assistance.

REFERENCES

1. D.J. Fitzgerald and A.S. Grove, Surf. Sci. **9**, 347 (1968).
2. E. Arnold and G. Abowitz, Appl. Phys. Lett. **9**, 344 (1966).
3. S.T. Hsu, D.J. Fitzgerald and A.S. Grove, Appl. Phys. Lett. **12**, 287 (1968).
4. D.M. Brown and P.V. Gray, J. Electrochem. Soc. **115**, 760 (1968).
5. P.V. Gray and D.M. Brown, Appl. Phys. Lett. **8**, 31 (1966).
6. P. Balk, Spring Meeting of the Electrochemical Society, San Francisco, CA, Abstract 109 (1965).
7. K.H. Zaininger, IEEE Trans. Nucl. Sci. **NS-13**, 237 (1966).
8. K.H. Zaininger and A.G. Holmes-Siedle, RCA Rev. **28**, 208 (1967).
9. D. Schmidt and G. Dorda, Elect. Lett. **17**, 761 (1981).
10. T.H. Ning, Solid State Electron. **21**, 273 (1978).
11. D.R. Young (this conference).
12. D.R. Young, E.A. Irene, D.J. DiMaria, R.F. DeKeersmaeker and H.Z. Massoud, J. Appl. Phys. **50**, 6366 (1979).
13. R.A. Gdula, J. Electrochem. Soc. **123**, 42 (1976).
14. P.J. Caplan, E.H. Poindexter, B.E. Deal and R.R. Razouk, J. Appl. Phys. **50**, 5847 (1979).
15. E.H. Poindexter, P.J. Caplan, B.E. Deal and R.R. Razouk, J. Appl. Phys. **52**, 879 (1981).
16. Y. Nishi, Jpn. J. Appl. Phys. **5**, 333 (1966).
17. Y. Nishi, Jpn. J. Appl. Phys. **10**, 52 (1971).
18. K.L. Brower, Appl. Phys. Lett. **43**, 1111 (1983).
19. P.M. Lenahan, K.L. Brower, P.V. Dressendorfer and W.C. Johnson, IEEE Trans. Nucl. Sci. **NS-28**, 4105 (1981).
20. P.M. Lenahan and P.V. Dressendorfer, J. Appl. Phys. **54**, 1457 (1983).
21. E.H. Poindexter, P.J. Caplan, J.J. Finnegan, N.M. Johnson, D.K. Biegelsen and M.D. Moyer, *The Physics of MOS Insulators*, ed. by G. Lucovsky et al. (Pergamon, NY 1980) pp. 326-330.
22. P.M. Lenahan and P.V. Dressendorfer, Appl. Phys. Lett. **41**, 542 (1982).
23. N.M. Johnson, D.K. Biegelsen, M.D. Moyer, S.T. Chang, E.H. Poindexter and P.J. Caplan, Appl. Phys. Lett. **43**, 563 (1983).
24. B. Henderson, Appl. Phys. Lett. **44**, 228 (1984).
25. K.L. Ngui and C.T. White, J. Appl. Phys. **52**, 320 (1981).
26. R.B. Laughlin, J.D. Joannopoulos and D.J. Chadi, Phys. Rev. **B21**, 5733 (1980).
27. T. Sakari and T. Sugano, J. Appl. Phys. **52**, 2889 (1981).
28. A. Goetzberger, V. Heine and E.H. Nicollian, Appl. Phys. Lett. **12**, 95 (1968).
29. C.T. Sah, IEEE Trans. Nucl. Sci. **NS-23**, 1563 (1976).
30. J. Singh, A. Madhukar, Appl. Phys. Lett. **38**, 884 (1981).
31. S.T. Chang, J.K. Wu and S.A. Lyon (to be published).
32. C.C. Chang, Ph.D. dissertation, Princeton University, 1976.

33. J.J. Clement, Ph.D. dissertation, Princeton University, 1977.
34. C.S. Jenq, Ph.D. dissertation, Princeton University, 1977.
35. G.J. Hu and W.C. Johnson, J. Appl. Phys. **54**, 1441 (1983).
36. G.J. Hu and W.C. Johnson, Appl. Phys. Lett. **36**, 590 (1980).
37. P.V. Gray, Proc. IEEE **57**, 1543 (1969).
38. M. Lenzlinger and E.H. Snow, J. Appl. Phys. **40**, 278 (1969).
39. S.T. Chang and S.A. Lyon (to be published).
40. G.J. Hu, Ph.D. dissertation, Princeton University, 1979.
41. J.K. Wu, S.A. Lyon and W.C. Johnson, Appl. Phys. Lett. **42**, 585 (1983).
42. P.M. Lenahan, Bull. Amer. Phys. Soc. **28**, 247 (1983).
43. P.S. Winokur, H.E. Boesch Jr., J.M. McGarrity and F.B. McLean, IEEE Trans. Nucl. Sci. **NS-24**, 2113 (1977).
44. P.S. Winokur, H.E. Boesch Jr., J.M. McGarrity and F.B. McLean, J. Appl. Phys. **50**, 3492 (1979).
45. S.K. Lai, Appl. Phys. Lett. **39**, 58 (1981).
46. A.M. Goodman, Phys. Rev. **144**, 588 (1966).
47. S. Pang, S.A. Lyon and W.C. Johnson, Appl. Phys. Lett. **40**, 709 (1982).
48. S. Pang, S.A. Lyon and W.C. Johnson, *The Physics of MOS Insulators*, ed. by G. Lucovsky et al. (Pergamon, NY, 1980) pp. 285-289.
49. V. Zekeriya and T.P. Ma, Appl. Phys. Lett. **43**, 95 (1983).
50. T.P. Ma, Appl. Phys. Lett. **27**, 615 (1975).
51. R.E. Mikawa and P.M. Lenahan (to be published - presented at IEEE Radiation Effects Conference, Colorado Springs, CO, 1984).
52. E.H. Nicollian and C.N. Berglund, J. Appl. Phys. **41**, 3052 (1970).
53. E.H. Nicollian, C.N. Berglund, P.F. Schmidt and J.M. Andrews, J. Appl. Phys. **42**, 5654 (1971).
54. T. Sunaga, S.A. Lyon and W.C. Johnson, Appl. Phys. Lett. **40**, 810 (1982).
55. M. Kuhn, Solid State Electron. **13**, 873 (1970).
56. J.K. Wu, Ph.D. dissertation, Princeton University, 1983.
57. J.Y. Teng and K. Hess, J. Appl. Phys. **54**, 5145 (1984).

THE DIFFUSION OF ION-IMPLANTED BORON IN SILICON DIOXIDE

Jacob Ng, James F. Gibbons and Thomas Sigmon

Stanford Electronics Laboratories, Stanford University
Stanford, CA 94305

ABSTRACT

In this work we report a detailed SIMS analysis of the diffusion of ion implanted B in SiO_2. Using the joined half Gaussian approximation to the initial B profile the diffusion of B has been modeled for temperatures of 1050° to 1200°C. A fast diffusing tail has been observed for diffusion temperatures below 1050°C. A method is suggested to minimize this "tail" diffusions.

The diffusion of boron in silicon has been extensively studied and excellent models have been developed to explain the phenomenon. However, the diffusion of boron in silicon dioxide is not as well understood nor as thoroughly studied. Values for the diffusivity of B in silicon dioxide show a great deal of variation[1-9]. Also, most of the previous experimental anaylsis used the mathematical model developed by Sah et al.[10] to derive the diffusivity of B in silicon dioxide. In this model, the diffusivities are calculated by solving a two boundary value problem with the assumption that the diffusivities of the dopant are constant in both the silicon and silicon dioxide. The diffusivities of boron in silicon dioxide are then determined by measuring the surface concentration or junction depth in the silicon, rather than the actual concentration profile of boron in silicon dioxide.

In this work, two different boron doses (1 x 10^{15} cm^{-2} and 2 x 10^{14} cm^{-2}) are ion-implanted at 30 keV into both dry and wet silicon dioxide of approximately 4000 Å thickness. The boron profiles before and after annealing are then measured by SIMS (secondary ion mass spectroscopy). The diffusion of boron was studied at temperatures between 1000°C to 1200°C in 50°C intervals.

Although SIMS is a very powerful tool for depth profiling, it is nevertheless incapable of giving quantitative results accurate enough for calculating the diffusivities of impurities in insulating materials. However, a novel method developed in this work of preparing the samples and manipulating the SIMS data has eliminated most of the instrument errors. The following describes the problems encountered in the analyses and the methods used to solve them.

Layers of SiO_2, 4000 Å thick, were grown on Si <100> surfaces at 1000°C in steam. The Si (n-type) wafers are cleaned by standard cleaning procedures before oxide growth. Boron was ion-implanted at 30 keV to a dose of 10^{15} cm^{-2} into the oxides. The thickness of the oxide was accurately known. Following implantation a layer of Si_3N_4 350 Å in thickness, is deposited onto the oxide by LPCVD at 625°C (Fig. 1). Similar samples with a lower implant dose (2 x $10^{14} cm^{-2}$) were prepared by the same methods.

The thin layer of Si_3N_4 prevents outdiffusion of the boron from the SiO_2 during heat treatment. The SiO_2/Si boundary can be considered to be infinitely far away for the mathematical analysis since the implanted energy is low.

Both $^{11}B^+$ and $^{30}Si^+$ (or $^{18}O^+$) SIMS signals are followed so that the abrupt change of the $^{30}Si^+$ (or $^{18}O^+$) signal due to the difference in sputtering yields of the $^{30}Si^+$ ions in SiO_2/Si will indicate where this interface is located. Since the thickness of the oxide is known, one can easily obtain a depth profile of the $^{11}B^+$ signal from the raw data. Since the diffusion barrier (Si_3N_4) prevents dopant loss during annealing, the dopant profile for each sample can be integrated and scaled to give the correct dose.

For the lower dose case (concentration peak at $2 \times 10^{19}/cm^3$), the electron multiplier was used as the detector (instead of the Faraday cup) to attain sufficient dynamic range of the B signal for accurate analysis. The $^{30}Si^+$ signal will saturate in the Cameca IMS3F in this mode, therefore the $^{18}O^+$ signal is used. Samples with oxides grown at 1100°C were prepared and analyzed in the exact manner.

PROBLEMS INVOLVED IN DOING SIMS ANALYSES ON SiO_2

When the sample is bombarded by a charged particle beam, the primary ions appear as a current flow to the sample. Secondary electrons will be emitted from or attached to the solid surface. If the algebraic sum of these individual currents is non-zero, there will be local surface potential inhomogeneities (perhaps fluctuating with time) which will also distort the extraction field. This situation can arise with bulk insulators, dielectric film layers or insulating microparticles such as dust or dirt. Extraction field alterations can occur in either dynamic or static equilibria. This will cause a dispersing effect on the net angular and energy distribution as seen by the mass spectrometer. These effects cause a severe problem in this analysis. One way to minimize this type of effect is to sputter the sample very slowly to avoid the charging problem and repeat the analysis whenever such phenomenon occur. Such analysis is very time consuming.

When O_2^+ primary ions bombard the sample, they produce a sample voltage variation which can be automatically corrected in the range 0 to 120 V. When high resistivity samples, like SiO_2, are analyzed, the voltage variation is higher than 120 V. In order to compensate for the charge accumulated at the surface, an electron flood gun can be used. The primary beam impact area must be entirely covered with the electron beam. As the impact position varies according to the primary ion energy, the electron beam position is adjusted by changing the electron energy between 0 keV and 2.5 keV. This corresponds to 5 keV to 7.5 keV net energy at the sample surface for this instrument. The electron beam current is adjusted by tuning the filament current.

Even using the electron gun, there is still a slight variation of the sputtered ion yield as seen by the mass spectrometer. This is probably due to local variation in the surface potential of the sample. The result is shown in Fig. 2. The variation in sputtered yield is shown by the variation of the $^{18}O^+$ signal which should be constant. Since the sputtering yield variations of the $^{18}O^+$ and $^{11}B^+$ ions should be the same, the variation of the $^{18}O^+$ signal gives a clue to how the $^{11}B^+$ signal actually varies. To obtain profiles accurate enough for this work, the $^{11}B^+$ signal is divided by the $^{18}O^+$ signal in each cycle before the profile is integrated and scaled to give the right concentration profile. The difference between profiles with and without utilizing such a procedure is illustrated in Fig. 3. This technique is quite useful in obtaining accurate SIMS profiles of boron in silicon dioxide.

EXPERIMENTAL RESULTS

The diffusivities are derived from both wet and dry oxides implanted at 30 keV with two doses: 10^{15} cm^{-2} and 2×10^{14} cm^{-2}. The diffusivities are estimated from 1050°C to 1200°C. The diffusivities of boron in both wet and dry silicon dioxide are derived by fitting the calculated profiles (from the joined half-Gaussian approximation and the original profile) to the profile obtained from the SIMS data.

Figure 4 shows SIMS data for both the unannealed samples and samples annealed at 1200°C for 16 and 40 hours, respectively and the joined half-Gaussian approximation of the unannealed profile as well as the diffusion profiles calculated from it. The diffusion profiles are calculated for $D_{ox} = 4.8 \times 10^{-17}$ cm^2/sec. These plots indicate that the diffusivity of boron at 1200°C is concentration independent, since a single value of D_{ox} results in profiles which fit the data over 4 orders of magnitude. The profiles for the lower dose are shown in Fig. 5. The result also indicates that boron diffusion at 1200°C is concentration independent.

The diffusivities of boron were determined for both dry and wet silicon dioxide and the results are plotted in Fig. 6. The pre-exponential factors and activation energies are tabulated in Table I. The activation energies vary from 5.13 eV to 5.8 eV (average = 5.57 eV). Figure 7 compares the results (wet, high dose) in this work with Horiuchi's results. The plots indicate that the diffusivity of boron is strongly dependent upon how the boron is introduced into the silicon dioxide. This work suggests that a change in the SUPREM modeling program is required to correctly model the diffusion of ion-implanted boron in silicon dioxide.

Table I: D_o's and E's for the various oxides

Features	D_o	E(eV)
High dose, wet	4846	5.8
High dose, wet	2178	5.75
High dose, dry	968	5.58
Low dose, dry	22	5.136

ENHANCED TAIL DIFFUSION

It was found that in both high dose and low dose cases, for annealing temperatures between 1000°C and 1100°C, a B tail diffusion occurs. This phenomenon is especially obvious for samples annealed at 1100°C. Because of this tail diffusion, the normal technique used to determine the diffusivity is somewhat misleading. The tail diffusion can cause a severe limitation in silicon gate technology. It is essential to characterize and control this type of diffusion.

HIGH DOSE CASE

Figure 8 shows the SIMS profiles for the annealed sample and sample annealed at 40 hours at 1050°C and 1100°C, respectively. In order to ascertain that B did diffuse into silicon at 1100°C and the results seen were not due to a possible ion-mixing or any other SIMS artifact, a sample after being annealed at 1100°C for 40 hours was plasma etched and then etched in buffered oxide to remove the nitride and oxide on top of the silicon. $^{11}B^+$ was detected near the silicon surface. The result confirms that some boron did diffuse into the silicon after 40 hours of annealing at 1100°C. Such tail diffusion into silicon could affect the threshold voltage of MOS devices.

It has been suggested[11] that wormholes exist in silicon dioxide. If wormholes really exist in silicon dioxide, interstitial B atoms may be able to diffuse very fast through them and give rise to a tail diffusion. If we anneal the oxide at 1200°C plastic flow would close up these worm holes and the tail diffusion shouldn't occur if the oxide was annealed later at lower temperatures. To determine whether pre-annealing at 1200°C did change the diffusive properties of the oxide, a sample was first annealed at 1200°C for 24 hours and then at 1100°C for 45 hours. The SIMS profiles of these samples indicate that pre-annealing at 1200°C does not stop tail diffusion at 1100°C.

From these results it seems as if some B has precipitated at high concentration and that the solubility of B may be a function of temperature. In order to verify this speculation, TEM was performed, however these results show no obvious sign of more than one phase being present.

LOW DOSE CASE

The anomalous diffusion phenomenon was not as obvious in the low dose case as in the high dose. Still there is a tail diffusion for samples annealed between 1000°C and 1100°C. This effect is most obvious for samples annealed at 1100°C.

EFFECT OF PRE-ANNEALING AT 600°C FOR TWO HOURS BEFORE HIGH TEMPERATURE ANNEALING

We discovered that pre-annealing at 600°C for two hours before the high temperature annealing (in the range between 1000°C and 1100°C) will reduce the tail diffusion for the low dose case. These results are shown in Fig. 9a,b. This effect may be significant for device fabrication and process modeling. This effect is not as obvious in the high dose case.

SUMMARY

The electron flood gun is necessary but not sufficient to obtain accurate SIMS profiles in insulating materials. An approach involving the following steps were developed for this work:
  1. The $^{11}B^+$ signal is divided by the $^{18}O^+$ signal in order to minimize the distortion of the profile due to the charging effect.
  2. The profile is then integrated and scaled to the right concentration profile corresponding to the implanted dose.

The diffusion of B in silicon dioxide has the following features:
  1. For annealing temperatures between 1150°C and 1200°C, standard diffusion occurs. Between 1150°C and 1100°C, an enhanced tail diffusion appears. The concentration of the tail depends upon the implanted dose. If the same phenomenon occurs for other techniques of B introduction, it could lead to a significant error if Sah's model were used to analyze the result.
  2. Pre-annealing the samples for two hours before high temperature annealing (in the range between 1000°C and 1100°C) will reduce the tail diffusion for the low dose case.
  3. The diffusivity of B in silicon dioxide is strongly dependent upon the experimental conditions.
  4. The D_p's are concentration independent, since a single value of D_p gives profiles which fits the SIMS data over 4 orders of magnitude. The D_p's seem to be weakly dependent on the dose.

REFERENCES

1. A. S. Grove, O. Leistiko, Jr. and C. T. Sah, J. Appl. Phys., 35, 2965 (1964).
2. M. L. Barry and P. Olofsen, J. Electrochem. Soc., 116, 854 (1969).
3. O. Schwenker, J. Electrochem. Soc., 118, 313 (1971).
4. D. M. Brown and P. R. Kennicott, J. Electrochem. Soc., 118, 293 (1971).
5. S. Horiuchi and J. Yamaguchi, Japan J. Appl. Phys. 1, 314 (1962).
6. K. V. Anand et al., J. Phys. D. App. Phys. 4, 1722 (1971).
7. P. R. Wilson, Sol. St. El. 15, 961 (1972).
8. S. P. Mukherjee et al. Thin Sol. Films 14, 299 (1972).
9. K. Shimakura et al., Sol. St. El. 18, 991 (1975).
10. C. T. Sah et al., J. Phys. Chem. Solids, 11, 288 (1959).
11. W. A. Pliskin, P. D. Davidse, H. S. Lehman and L. I. Maissel, IBM J. Res. Dev. 11, 461 (1967).

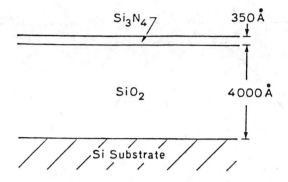

Fig. 1   Structure of the samples prepared for SIMS analyses.

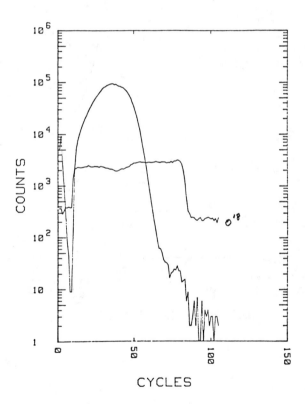

Fig. 2   SIMS profile of a sample illustrating a slight charging effect.

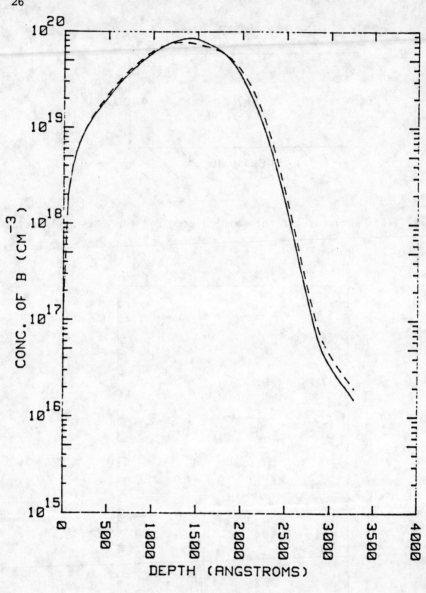

Fig. 3 B profiles with and without normalizing with respect to the $^{18}O^+$ signal.
_____ normalized profile
_ _ _ unnormalized profile

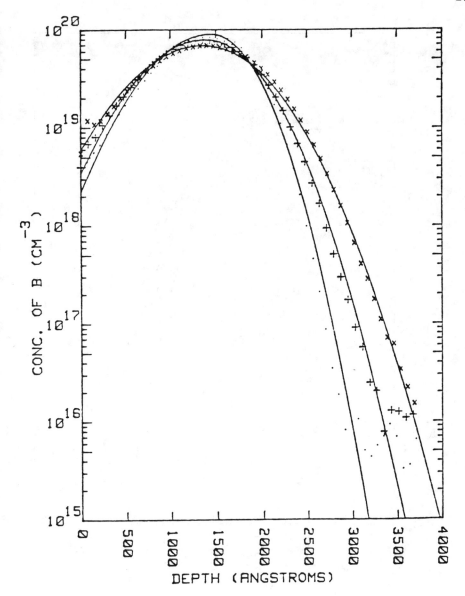

Fig. 4 Calculated and actual SIMS profiles for B diffusion in wet silicon dioxide;
. . . unannealed,
+ + + annealed at 1200°C for 16 hours,
x x x annealed at 1200°C for 40 hours.
The diffusion profiles are calculated with $D_{ox} = 4.8 \times 10^{-17}$ cm^2/sec from the joined half-Gaussian approximation of the unannealed profile for a dose of 10^{15} cm^{-2}.

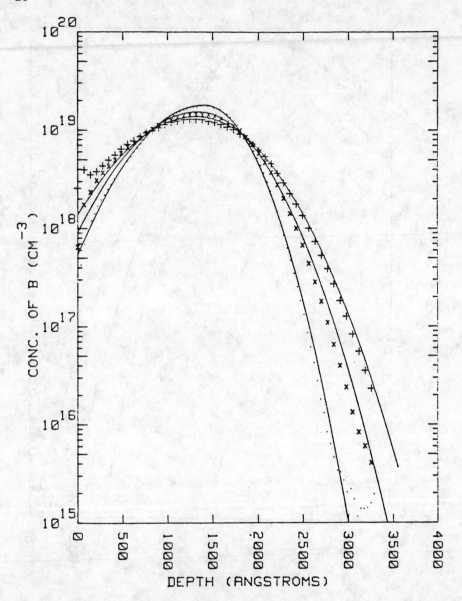

Fig. 5 Calculated and actual SIMS profiles of B in silicon dioxide;
. . . unannealed
x x x annealed at 1200°C for 16 hours
+ + + annealed at 1200°C for 32 hours
The diffusion profiles are calculated with $D_{ox} = 6.0 \times 10^{-17}$ cm^2/sec from the joined half-Gaussian approximation of the unannealed profile for a dose of 2×10^{14} cm^{-2}.

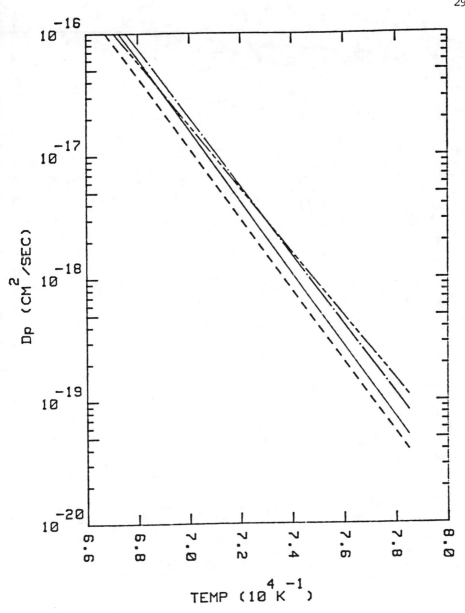

Fig. 6 Least Squares fits of D_p's of B in silicon dioxide.
———— low dose, wet
– – – – high does, wet
▬ – – ▬ low dose, dry
· — · — · high dose, dry

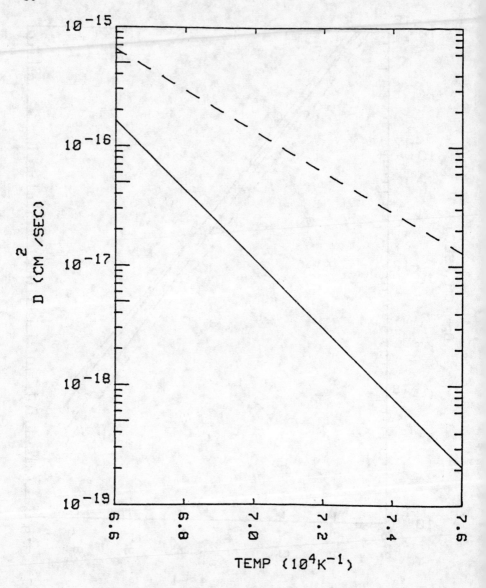

Fig. 7 Comparison of the result in this work and Horiuchi's result.

——— this work, ion-implanted B in SiO_2 (wet, high dose)

- - - Horiuchi's work (wet, gaseous B)

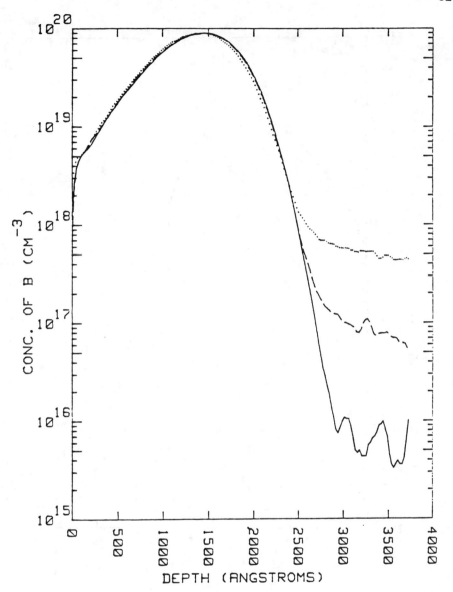

Fig. 8   SIMS profiled for B diffusion in wet silicon dioxide:
　　　　───── unannealed
　　　　── ── annealed at 1050°C for 40 hours
　　　　..... annealed at 1100°C for 40 hours

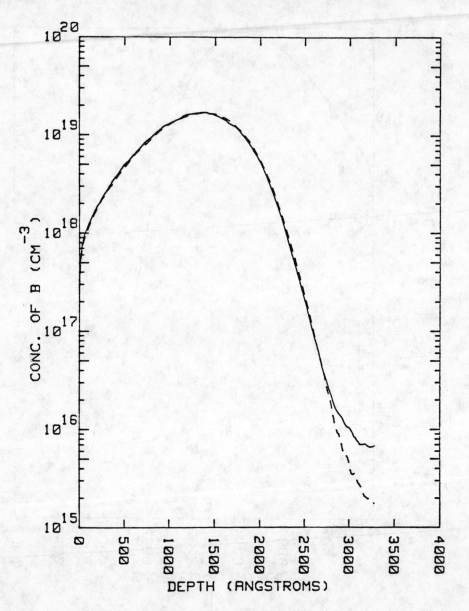

Fig. 9(a) SIMS profiles for B diffusion in dry silicon dioxide for samples annealed at 1100°C for 48 hours; dose = $2 \times 10^{-14} cm^{-2}$.
——— without 600°C pre-annealing
- - - with 600°C pre-annealing

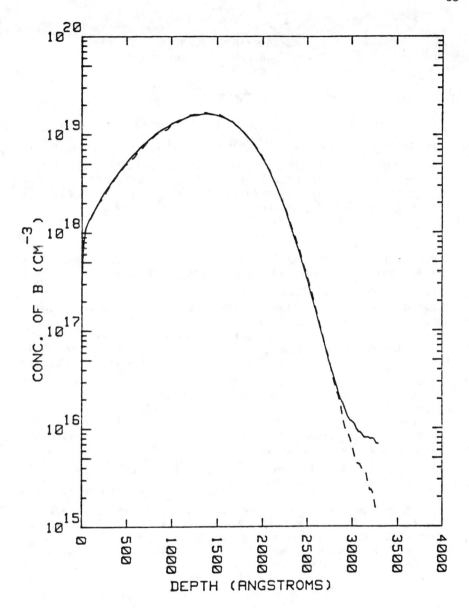

Fig. 9(b) SIMS profiles of B in dry silicon dioxide for samples annealed at 1100°C for 80 hours; dose = $2 \times 10^{-14} \text{cm}^{-2}$
———— without 600°C pre-annealing
- - - - with 600°C pre-annnealing

EFFECT OF HYDROGEN ON Si-SiO$_2$ STRUCTURES: AN ESR STUDY

W. E. Carlos and H. L. Hughes
Naval Research Laboratory
Washington, D.C. 20375

ABSTRACT

Implantation of hydrogen into Si-SiO$_2$ structures is seen to increase the number of E' centers in the SiO$_2$ induced by ionizing irradiation. Correlation of the concentration of these centers with the ESR signal due to conduction electrons in the accumulation layer indicates that these centers are related to the positive interface charge density.

INTRODUCTION

Hydrogen has long been recognized as an important impurity in SiO$_2$ films for Si MOS devices.[1] It is generally accepted that small quantities of hydrogen reduce the interface charge density by tying up dangling bonds or by relieving strained bonds. SIMS measurements support this by revealing a buildup of hydrogen at the interface.[2] Water diffusion experiments by Feigl and coworkers[3] show a correlation between water and electron traps in the bulk of the oxide. Infrared measurements reveal SiOH, intersial H$_2$O and to a lesser extent Si-H bonding configurations in films grown by steam oxidation, films into which H$_2$O had been diffused and in hydrogen implanted films.[4,5]

Electron spin resonance (ESR) has proved to be a valuable tool for probing defects in bulk amorphous SiO$_2$[6] and over the last decade has provided many insights into the defect structure of thermal SiO$_2$ films and Si-SiO$_2$ interfaces.[7] In bulk SiO$_2$ there are a number of defects which have been associated with hydrogen through deuterium substitution experiments,[8] although in many cases no structure has been established. In addition, the density and annealing behavior of other bulk defects such as the E' center (a trivalent Si atom) or the peroxy radical (Si-O-O-) are affected by the OH content of the silica.[9] In light of the successful use of ESR for studying hydrogen in bulk SiO$_2$ and in other problems of the Si-SiO$_2$ interface we undertake the study of hydrogen in SiO$_2$ films using ESR as the primary probe.

EXPERIMENTAL PROCEDURES

Samples for the work were all prepared from phosphorous doped (~30 Ωcm), silicon with (100) orientation. The silicon wafers were polished on both sides to double the oxide area. Dry thermal oxides were grown at 1000°C to a thickness of 1000 Å. A 1500 Å Si$_3$N$_4$ cap was deposited over the oxide and the samples were implanted with hydrogen at an energy of 30 keV. The samples were then annealed at 400°C for 2 hours followed by 300°C for 1 hour and the cap was then removed. A thin (~500 Å) layer of aluminum was evaporated over the SiO_2 and leads were bonded to it and the silicon substrate to permit biasing either during irradiation or during measurements. The total area of the devices was one square centimeter. ESR measurements were made using a standard Varian spectrometer operating at 9.3 GHz. Data were taken with the sample held at 15K using an Air Products Helitran flow dewar. At these low temperatures many resonances due to isolated centers such as E' are completely saturated. In such cases it is necessary to employ rapid passage techniques to achieve any detectable signal.[10] In this work the second harmonic of the absorption signal, 90° out of phase with the field modulation, is detected. This particular method has been applied to the problem of slowly tumbling molecules in chemistry.[11] In general it provides superior signal to noise and line resolution to the dispersion signal commonly detected in rapid passage experiments. As in most rapid passage experiments the lineshape is approximately that of the undifferentiated absorbtion signal.

In order to generate electron-hole pairs in the oxide and populate some of the traps, the samples were irradiated using a copper x-ray tube operating at 100 kV. All irradiations were done at room temperature with the metal biased at +10 V.

RESULTS AND DISCUSSION

No ESR signal due to oxide or the interface was detectable prior to irradiation. However, after irradiation a signal is seen in those samples which had been implanted. After signal averaging a similar signal is resolved in the unimplanted material. In Fig. 1 the spectra for samples with implantation doses of $10^{15}/cm^2$ and $10^{16}/cm^2$ are shown along with the spectra for an unimplanted sample. The two strong resonances on either side are due to neutral phosphorous in the bulk and are of no concern here. The line at the center is due to E' centers (trivalent silicon backbonded to three oxygen atoms). No center directly related to hydrogen is seen and there is little evidence for any oxygen hole centers although the low field phosphorous line would partially obscure such a line.

In addition to the E' centers which are strongly saturated, an unsaturated line is seen at $g \cong 1.9997$, also shown in Fig. 1. This line is due to conduction electrons in the accumulation layer formed when positive charge accumulates in the oxide. A conduction electron spin resonance (CESR) is easily distinguished from a resonance due to an isolated paramagnetic center in that its intensity is not Curie-like,[12] i.e. it is not a strong function of temperature. The magnetic susceptibility of such a resonance is not simply proportional to the number of electrons in the accumulation region but rather is proportional to the density of states at Fermi level and in the case of an accumulation layer will depend on the bending of the bands at the surface.

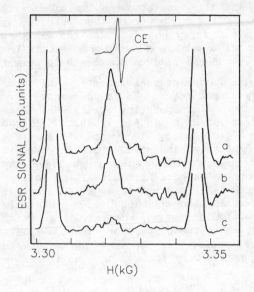

Fig. 1. The central portion of the ESR spectrum of three samples; (a) implanted with $10^{16}H/cm^2$, (b) implanted with $10^{15}H/cm^2$ and (c) unimplanted. All have been x-irradiated for 15 min. Also shown is the conduction electron resonance (CE) from the sample implanted with $10^{16}H/cm^2$.

In order to better understand the relationship between these signals we consider the rate at which they are produced by x-irradiation as shown in Fig. 2. The data shown there were taken on a sample implanted to a dose of $10^{16}/cm^2$. We first focus on the relationship between the E' concentration and irradiation time (t). Galeener and coworkers[13] have fit similar data for bulk SiO_2 to a relation of the form

$$N(E') = At + B(1-e^{-t/\tau}) \quad , \qquad (1)$$

arguing that the first term is due to the actual creation of centers and that the second term is due to population of existing sites by photoexcited carriers. On the other hand Winokur and coworkers[14] have fit the interface charge density in irradiated MOS devices to a $t^{2/3}$ dependence. This power law dependence implies that

$$\frac{dN}{dt} = \frac{\alpha}{\sqrt{N}} \quad , \qquad (2)$$

and we see no simple physical model which would yield these kinetics. Best fits to both Eq. (1) and a power law dependence are shown in Fig. 2. As can be seen there is very little difference between the two curves over this limited a range of data. Indeed, the data given in either ref. 13 or 14 can be reasonably fit by either functional form. While we are inclined to believe that Eq. (1) more accurately models the kinetics simply because it is backed by a simple physical picture, the resolution must await data over a wider range of irradiation times.

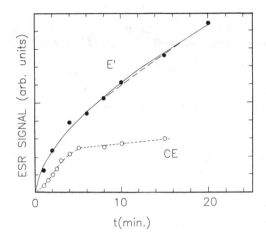

Fig. 2. The ESR signal as a function of irradiation time for the E' center and the conduction electrons (CE). The solid line is a best fit to equation (1), while the dashed line is a fit to a power law. The line through the CE data is merely an aid to the eye.

Also shown in Fig. 2 is the CESR as a function of irradiation. We see that the dose dependence is quite similar to that of E' center except that at high doses (t>5 min) the CESR levels off while the E' signal continues to increase linearly. This can be best explained by assuming that there are two different distributions of E' centers in these oxides. The first is a concentration near the interface which by trapping holes contributes to the interface charge and strongly affects the surface potential and hence the electron concentration in the accumulation layer. The remaining centers are more uniformly distributed in the oxide and do not have as strong an effect on the surface potential. Results for the unimplanted or more lightly implanted samples show a more linear dependence of the E' center concentration on irradiation time, although definitive statements about those samples are more difficult owing to the low concentration of such centers in those films.

The possible role of the E' center as a site for positive interface charge merits further discussion. The resonant field of a site actually at the physical interface such as a P_B center (a trivalent Si backbonded to three other silicon atoms) would be a function of the angle between the applied magnetic field and the surface and no such dependence is seen here. However, a site within about 50 Å of the surface could still be in intimate contact with the semiconductor via a simple tunneling mechanism. If such a site were located more than one or two interatomic distances from the interface, one would not expect to see the symmetry of the surface reflected in the ESR spectrum. Further evidence for this picture is seen in the room temperature annealing behavior of the E' resonance and the accumulation layer resonance. After seveal hours to a day the CESR is no longer detectable while the E' signal is significantly decreased (a factor of ~2/3 for the data shown in Fig. 2). Both signals are refreshed upon re-irradiation.

Although the model of an E' related hole trap being responsible for interface traps in the hydrogen rich oxide fits the data reasonably well, the microscopic nature of the process is not easily resolvable. The Si atom which has an unpaired electron giving rise to the ESR signal is neutral. In the case of an Si-SiO$_2$ interface a simple picture of the trapping mechanism would be a strained Si-Si bond which after trapping a hole breaks into two trivalent silicon atoms, one positively charged and the other paramagnetic. Infrared data[4] indicate that Si-OH units occur in pairs or small clusters. We speculate that such a pair could trap a hole by the following reaction

$$2 \text{ Si-OH} + h^+ \rightarrow H_2O + E' + \text{Si-O}^+ \quad . \tag{3}$$

This would account for the concurrent appearance of positive interface charge and the paramagnetic E' center. Furthermore, this pair could later trap an electron,

$$\text{Si-O}^+ + E' + e^- \rightarrow \text{Si-O-Si} \tag{4}$$

thereby eliminating both the charge and the paramagnetism.

CONCLUSIONS

The number of E' ESR centers is seen to increase with increasing hydrogen content in the SiO$_2$ film. From the correlation between the E' centers and the conduction electron spin resonance from the accumulation layer in the silicon we conclude that the E' centers do play a role in the buildup of positive interface charge. The details of the role of the hydrogen in this process are not clear, however.

REFERENCES

1. B.E. Deal, E.L. MacKenna and P.L. Castro, J. Electrochem. Soc. 116, 997 (1969); P. Balk, Extended Abstracts, Electronics Division, Electrochem. Soc. 14, 237 (1965); K.H. Beckmann and N.J. Harrick, J. Electrochem. Soc., 118, 614 (1971); A.G. Revesz, IEEE Trans. Nucl. Sci. NS-18, 113 (1971).
2. N.M. Johnson, D.K. Biegelsen, M.D. Moyer, V.R. Deline and C.A. Evans, Jr., Appl. Phys. Lett. 38, 995 (1981).
3. F.J. Feigl, D.R. Young, D.J. DiMaria, S. Lai and J. Calise, J. Appl. Phys. 52, 5665 (1981).
4. A. Harstein and D.R. Young, Appl. Phys. Lett. 38, 631 (1981).
5. E.D. Palik, R.T. Holm, A. Stella and H.L. Hughes, J. Appl. Phys. 53, 8454 (1982).
6. For a review see D.L. Griscom, J. Non-Crystalline Solids 31, 241 (1978).
7. E.H. Poindexter, P.J. Caplan, B.E. Deal and R.R. Razouk, J. Appl. Phys. 52, 879 (1981); K.L. Brower, Appl. Phys. Lett. 43, 1111 (1983).
8. J. Vitko, Jr., J. Appl. Physics 49, 5530 (1978).
9. D.L. Griscom, to be published.
10. For detailed review of passage effects see M. Weger, Bell Sys. Tech. Jour. 39, 1013 (1960).
11. L.R. Dalton, B.H. Robinson, L.A. Dalton and P. Coffey, Advances in Mag. Res. (ed. J.S. Waugh) 8, 149 (Acad. Press, N.Y.), 1976.
12. Y. Yafet, Solid State Physics 14, 1 (1960).
13. F.L. Galeener and T.C. Mikkelsen, Jr., to be published.
14. P.S. Winokur, H.E. Boesch, Jr., J.M. McGarrity and F.B. McLean, IEEE Trans. on Nuc. Sci. NS-24, 2113 (1977).

Part II. Surfaces and Interfaces

EFFECT OF ANNEALING CHARGE INJECTION AND ELECTRON BEAM IRRADIATION ON THE Si-SiO$_2$ INTERFACE BARRIER HEIGHTS AND ON THE WORK FUNCTION DIFFERENCE IN MOS STRUCTURES

S. Krawczyk, M. Garrigues, T. Mrabeut
Laboratoire d'Electronique, Automatique et Mesures Electriques de l'Ecole
Centrale de Lyon, ERA (CNRS) n° 661
36, av. de Collongue - B.P. 163 - 69131 ECULLY Cedex - FRANCE

ABSTRACT

In this work we report the effects of postmetallization forming gas (F.G.) annealing, electron photoinjection and electron beam irradiation on the effective work function difference (Φ_{ms}) and effective interface barrier heights at the metal-insulator and semiconductor-insulator interfaces (Φ_m and Φ_s, respectively) in Al-SiO$_2$-Si (MOS) structures.

It is shown that the variations of Φ_{ms} are generally due to the simultaneous modifications of the barrier heights at both interfaces.

Special attention is paid to the phenomena taking place at the Si-SiO$_2$ interface.

INTRODUCTION

In recent years the phenomena affecting the apparent value of the metal-silicon work function difference in MOS structures have been subjected to extensive experimental study, with a two fold objective for VLSI:
(1) to control the initial value of the flatband voltage V_{FB};
(2) to explain and prevent V_{FB} instabilities for operating
   devices; of major importance are for example the hot
   carrier effects arising in short-channel devices.

The dispersion of Φ_{ms} values reported in the literature exceed 0.4eV for the Al-SiO$_2$-Si system. This has been explained by dipole layers located either at the Al-SiO$_2$[1] or Si-SiO$_2$[2] interfaces. Up to now, results have been based on Φ_{ms} measurements carried out by the method of Kar[3] which do not permit one to distinguish between the phenomena taking place at the Al-SiO$_2$ and Si-SiO$_2$ interfaces.

In this work, the effects of each interface have been separated through direct measurements of Φ_{ms} and of the barrier heights at the metal-SiO$_2$ (Φ_m) and Si-SiO$_2$ (Φ_s) interfaces using photoelectric techniques. It is shown that barrier heights at both interfaces can be affected by post-metal annealing and electron irradiation or internal photoinjection from Si into SiO$_2$. We have paid particular attention to the phenomena taking place at the Si-SiO$_2$ interface.

EXPERIMENTAL

Al-SiO$_2$-Si structures with 1000 Å thick oxide layers were used in this study. Wet thermal oxidation of <100> silicon wafers was performed at 950° in a quartz tube furnace previously subjected to an HCl cleaning procedure.

MOS capacitors were submitted to one of the following treatments or their combination:
  - annealing at 400°C in a forming gas (F.G.), (10% H$_2$, 90% N$_2$),
  - irradiation by a 25 keV electron beam performed in a S.E.M.,

- internal photoinjection from Si into SiO_2 at $\lambda = 260$ nm (current density 10^{-9} to 10^{-8} A/cm^2, $V_G = 10$ volts).

Φ_m, Φ_s as well as Φ_{ms} were measured independently by classical photoemission methods[4,5,6]. Interface state and charge densities were evaluated by C(V) and G(V) techniques.

EFFECT OF 400°C FORMING GAS ANNEALING

Fig. 1 shows the compiled results for the values of Φ_m and Φ_s as a function of annealing time. The error bars correspond to the dispersion over the set of samples investigated. Two effects determining the evolution of Φ_{ms} during the annealing are apparent:
(1) a steep decrease of Φ_s and associated dispersion in the first 20 min (−0.1V); for longer annealing times, Φ_s tends to a slightly higher value with a continuously increasing dispersion that can be due to the diffusion of extrinsic defects;
(2) an increase of Φ_m which reaches a maximum at $t \approx 30$ min (+0.15 eV). It is due to the direct $Al-SiO_2$ interaction and interdiffusion phenomena [7,8,9].

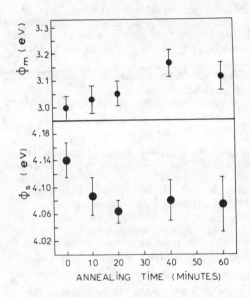

Fig. 1 : Effects of annealing time on the $Si-SiO_2$ (Φ_s) and $Al-SiO_2$ (Φ_m) interface barrier heights.

EFFECT OF ELECTRON IRRADIATION

Typical results for an irradiation dose of 10^{-3} C/cm^2 are summarized in Table I for F.G. post-metal annealed (400°C, 20 min) (PMA) and non post-metal annealed (NPMA) samples.

These results have been analyzed in detail elsewhere[10]. The variations of Φ_m are controlled by two phenomena: electron trapping and positive charge creation. The first increase of Φ_m just after irradiation is due to the predominant effect of electron trapping close to Al. Electron trapping is reversible and several days after irradiation Φ_m is usually lower than before irradiation as a result of the remaining positive charge.

	Before irradiation		Immediately after irradiation		Several days after irradiation	
	NPMA	PMA	NPMA	PMA	NPMA	PMA
ϕ_m (eV)	3	3.1	3.05	3.4	≃ 3	≃ 3
ϕ_s (eV)	4.1	< 4.1	4.3	4.15	4.3	4.15
ϕ_{ms} (eV)	-0.45	-0.25	direct measurement rather difficult		≃ -0.7	≃ -0.4

Table I : Effects of a 10^{-3} C/cm^{-2} electron-beam irradiation on Φ_m, Φ_s and Φ_{ms} for post-metal annealed (PMA) and non post-metal-annealed (NPMA) MOS structures.

EFFECT OF CARRIER PHOTOINJECTION INTO SiO$_2$

Data collected on several PMA samples show that in this case, injection results in a decrease of the Si-SiO$_2$ barrier height (Fig. 2). In the NPMA samples the effect is smaller or even a slight increase of Φ_s is observed. Some evidence concerning

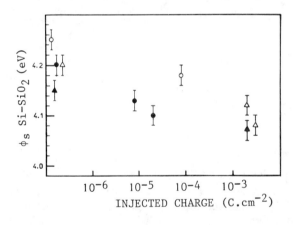

Fig. 2 : Effect of internal photoinjection of electrons from Si into SiO$_2$ for post-metal annealed MOS structures.

electron trapping phenomena and positive charge creation induced by charge injection at the Al-SiO$_2$ interface have been noticed, which can induce slight variations (± 50 meV) of Φ_m. However, these effects strongly depend on the processing conditions (contamination).

ON THE POSSIBLE CORRELATION BETWEEN THE INTERFACE PHENOMENA AND THE Si-SiO$_2$ BARRIER HEIGHT

The modifications of Φ_s observed in this work are summarized in Table II.
The apparent correlation between the intrinsic interface state density N_{it} and Φ_s due to F.G. annealing is shown in Fig. 3. It is weaker than the correlation between N_{it} and Φ_{ms} recently demonstrated by Razouk and Deal[2]. The difference is

Post-metal F.G. annealing	ϕ_S ↘ (about 0.1 eV)
Electron irradiation	NPMA samples : ϕ_S ↗ (0.1 ÷ 0.2 eV)
	PMA samples : ϕ_S ↗ (0.05 ÷ 0.1 eV)
Electron photoinjection	NPMA samples : (slight variation)
	PMA samples : ϕ_S ↘ (0.05 ÷ 0.15 eV)

Table II: Modifications of the Si-SiO$_2$ barrier height induced by F.G. annealing, electron irradiation and electron photoinjection.

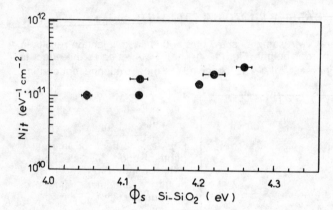

Fig. 3 : Correlation between the intrinsic interface state density (N_{it}) measured around mid-gap and the Si-SiO$_2$ interface barrier height Φ_S.

attributed to the variations of Φ_m which were not taken into account in[2]. In view of previous work[11,12], the decrease of Φ_s during annealing is deemed to be due to hydrogen chemisorption in the Si-SiO$_2$ interface region which can cancel previously existing dipoles located at this interface.

Assuming that dipoles of atomic dimensions ($\approx$ 0.2 nm) are responsible for the observed variations of Φ_s (0.1 eV), their density should be between 10^{13} and 10^{14} cm^{-2} and this value exceeds considerably simultaneous changes of N_{it} (5x10^{10} ÷ 2x10^{11} cm^{-2}).

Thus, the F.G. annealing must in fact affect about 10^{13}-10^{14} atoms or bonds at the interface, but only 0.01 - 1% of these structural modifications influences N_{it} in the middle of the gap. It must be noted that a strong density of extrinsic interface states observed accidently in some samples was not correlated with the corresponding value of Φ_s.

The above results indicate that the density of the broken, strained and/or deformed bonds which can attach hydrogen atoms and modify interface dipoles is much higher than the apparent H-compensated interface state density. This strongly suggests that intrinsic states arise from the statistical nature of the atomic configurations in the Si-SiO$_2$ interface region[14,16]. Experimental results[13,14,15] indicate the presence of a narrow ($\approx$ 2 atomic layers) disordered layer at the Si-SiO$_2$ interface, and show that fast surface states are located within the first 1.42 Å. This disorder can result in interface state bands located mainly outside the silicon band gap. N_{it} measurements reveal only band tails penetrating into the gap whereas $\Delta\Phi_s$ is sensitive to the overall density of disordered bonds.

The apparent correlation between Φ_s and N_{it} observed for intrinsic interface states (lower Φ_s - lower N_{it}) is no longer valid in the case of interface traps created by charge injection.

Charge injection results in an increase of N_{it} and at the same time in a decrease of Φ_s, particularly important in the PMA samples (Fig. 4). This effect cannot be explained on the basis of positive charge creation at the interface. In the light of a recent paper by Gale et al[17], it seems that the observed lowering of Φ_s may be due to the transport of hydrogen toward the Si-SiO$_2$ interface occurring during electron injection. The fact that the Φ_s lowering is more important in the F.G. annealed samples supports this hypothesis.

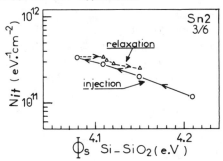

Fig.4 : Correlation between the interface state density (N_{it}) and Si-SiO$_2$ interface barrier height Φ_s as resulting from charge photoinjection.

In contrast, the predominant effect responsible for the Φ_s increase caused by electron irradiation at 25 keV is thought to be the breaking of Si-H and/or Si-OH bonds at the Si-SiO$_2$ interface. This effect may result in the observed significant modification of the dipole layer at this interface[10].

CONCLUSION

From direct photoemission measurements of the interface barrier height in MOS structures it was demonstrated that:
(i) the effective interface barrier height at the Si-SiO$_2$ interface (Φ_s) decreases after a 400°C F.G. annealing;
(ii) Electron beam irradiation results in an increase of Φ_s (particularly important in non-annealed samples);
(iii) Charge injection usually results in a decrease of Φ_s (particularly in the annealed samples);
(iv) there is an apparent correlation between the intrinsic interface states left by the material processing and the effective interface barrier height at the Si-SiO$_2$

interface (lower N_{it} - lower Φ_s). This correlation may be strongly modified by the presence of extrinsic states and is no longer valid for the case of interface states created by charge injection. Possible explanations of observed phenomena were given.

Our results also show that the variations of the barrier height at both interfaces must be considered in order to understand the modifications of the effective metal-silicon work function difference in MOS structures.

ACKNOWLEDGEMENTS

The authors are indepted to the CNET-Centre Norbert Segard in Meylan, France for providing oxidized silicon wafers. We gratefully acknowledge the e-beam exposures which were done by P. Frangville from centre de Microanalyse de l'Ecole Centrale de Lyon. We appreciate also many helpful discussions with P. Viktorovitch. This research was supported in part by the "Groupement Circuits Intégrés Silicium"(GCIS).

REFERENCES

1. T. W. Hicmott, J. Appl. Phys. 51, 4269 (1980).
2. R. R. Razouk and B. E. Deal, J. Electrochem. Soc. 129, 806 (1982).
3. S. Kar, Solid-State Electron. 18, 723 (1975).
4. R. Williams, Phys. Rev. A, 140, 569 (1965).
5. B. E. Deal, E. N. Snow and C. A. Mead, J. Phys. Chem. Solids 27, 1873 (1966).
6. S. Krawczyk, H. M. Przewlocki and A. Jakubowski, Revue Phys. Appl. 17, 473 (1982).
7. R. S. Bauer, R. Z. Bachrach and L. J. Brillson, Appl. Phys. Lett. 37, 1006 (1980).
8. P. M. Solomon and D. J. DiMaria, J. Appl. Phys. 52, 5867 (1981).
9. S. Krawczyk and T. Mrabeut, 14th European Solid-State Device Research Conference, Lille (France), Sept 10-14 (1984).
10. T. Mrabeut and S. Krawczyk, MRS-Europe Conference 1984 Meeting, Symposium III, Strasbourg (France), June 5-8 (1984).
11. A. K. Gaind and L. A. Kasprzak, Solid. State Electron. 22, 303 (1979).
12. Z. A. Weinberg and A. Harstein, J. Appl. Phys. 54, 2517 (1983).
13. D. E. Aspnes, J. B. Theeten and R. P. H. Chang, J. Vac. Sci. Technol. 16, 1374 (1979).
14. J. Singh and A. Madhukar, Appl. Phys. Lett. 38, 884 (1981).
15. M. Hamasaki, Solid-State Electron. 25, 205 (1982).
16. T. Sakurai and T. Sugano, J. Appl. Phys. 52, 2889 (1981).
17. R. Gale, F. J. Feigl, C. W. Magee and D. R. Young, J. Appl. Phys. 54, 6938 (1983).

DEPENDENCE OF Si-SiO$_2$ INTERFACE STRUCTURES ON CRYSTAL ORIENTATION AND OXIDATION CONDITIONS

Takeo Hattori, Masaaki Muto* and Toshihisa Suzuki
Musashi Institute of Technology, Setagaya-ku, Tokyo 158, Japan
*Presently, Fujitsu Laboratories, Atsugi, Kanagawa 243-01, Japan

Kikuo Yamabe
VLSI Research Center, Toshiba Corp., Saiwai-ku, Kawasaki 210, Japan

Hiroshi Yamauchi
Shimadzu Seisakusho Ltd., Nishinokyo, Nakagyo-ku, Kyoto 604, Japan

ABSTRACT

Based on the observation of crystal orientation dependence of Si 2p photoelectron spectra, the orientation dependence of interface structures and the distribution of intermediate oxidation states of Si in the oxide film were determined. The effects of oxidation condition on the structures of interfacial transition layer and SiO$_2$ were found.

INTRODUCTION

As a result of extreme decrease in the dimensions of metal-oxide semiconductor field-effect transistor device in recent years, the electronic states in Si-SiO$_2$ interfacial transition region and those in amorphous SiO$_2$ network play a vital role in device operation. Because of this, the chemical and physical structures of Si-SiO$_2$ interface and SiO$_2$ have been studied extensively by using various surface sensitive techniques. The existence of an abrupt interface,[1] the existence of disordered silicon layers next to the interface,[2] the existence of intermediate oxidation states of silicon,[3,4] structural changes in amorphous SiO$_2$ network near the interface,[5] crystal orientation dependence of Si-SiO$_2$ interface structures[6] and the structural origin of interface electronic states[7] were discovered or elucidated. It was also confirmed from electron microscopic observation that the interface is not flat on the atomic scale.[8]

EXPERIMENTAL RESULTS AND DISCUSSIONS

Nondestructive measurements with appropriately large probing depth are essential for the structural studies of abrupt interfaces such as Si-SiO$_2$ interface in order to avoid the effects of ion bombardment on the interface structures encountered in the conventional depth profiling studies.[9] The experimental details are the same as those described elsewhere.[6]

The spectrum a, b, and c shown by the solid lines in Fig.1 are Si 2p photoelectron spectra, excited by Mg Kα radiation, obtained for the oxide films formed in dry oxygen at 800°C on (100), (111), and (110) surface, respectively. The thickness of the silicon dioxide film for spectrum a, b, and c is 2.4, 2.1, and 2.3 nm, respectively. These spectra can be separated into three spectra originating from Si in silicon substrate, Si in silicon dioxide, and Si in an intermediate oxidation state. The spectra thus separated are shown by the dashed lines in Fig. 1. The difference for <100> and <111> orientations is observed in the intermediate oxidation state spectra. Namely, the spectrum observed at around 101.3 eV in spectrum a is not observed in spectrum b, while the spectrum observed at around 100.3 eV in

0094-243X/84/1220045-05 $3.00 Copyright 1984 American Institute of Physics

spectrum b is not observed in spectrum a. The spectrum observed at 101.3 eV and 100.3 eV agrees with previously assigned oxidation states, SiO and Si_2O, respectively.[5] The spectrum at around 102.6 eV is observed for three crystal orientations and agrees with previously assigned oxidation state, Si_2O_3.[5]

In order to investigate the intermediate oxidation states in more detail, the intermediate oxidation state spectra are separated by the least squares calculations with the following assumptions: (1) the intermediate oxidation states consist of three oxidation states, Si_2O, SiO, and Si_2O_3; (2) the line shape of each separated

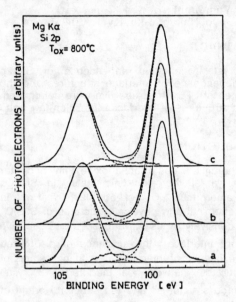

Fig. 1: Si 2p photoelectron spectra for three crystal orientations. Oxide films were formed in dry oxygen.

spectrum is the same as that for the chemically etched silicon. In Fig. 2 the intensities NI of three intermediate oxidation state spectra are shown as a function of silicon dioxide film thickness for the oxide films formed at 1000°C on (100), (111), and (110) surface, respectively. Here, the value of NI is normalized by the spectral intensity NS of substrate silicon for each spectrum. The value of (NI/NS) for SiO on (100) surface, that for Si_2O on (111) surface, and those for SiO and Si_2O on (110) surface weakly depends on the oxide film thickness and approaches a constant value with decreasing oxide film thickness respectively. On the other hand, the value of (NI/NS) of Si_2O_3 strongly depends on the oxide film thickness such that the values of (NI/NS) exhibit rapid decrease with decreasing thickness in addition to the slightly weaker thickness dependence as compared with that for (NO/NS).

These observations can be understood as follows with the assumptions that the distribution of the intermediate oxidation state in the oxides does not change with the oxide film thickness, in other words, the oxidation mechanism does not change with the oxide film thickness. If the intermediate oxides are located only at Si-SiO_2 interface, the value of (NI/NS) should be independent of the dioxide film thickness. Therefore, the weak thickness dependence of SiO spectral intensities for (100) surface, that of Si_2O spectral intensities for (111) surface, and those of SiO and Si_2O for (110) surface indicate their locations mostly at the Si-SiO_2 interface. On the other hand, the strong thickness dependence of Si_2O_3 spectral intensities for three crystal orientations indicates their distributions in the oxide film near the interface, but not at the interface. The amount of Si_2O_3 in the oxide film is less than 10% of the amount of SiO_2. These distributions of intermediate oxides in the oxide film can be also expected from the oxygen induced binding energy shifts observed as a function of the oxide film thickness.[10]

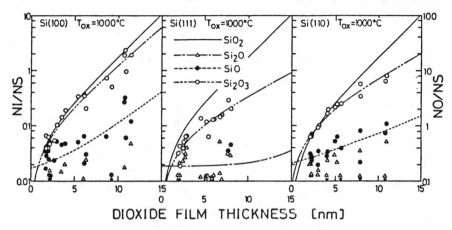

Fig. 2: Normalized spectral intensities for three intermediate oxidation states are shown as a function of silicon dioxide film thickness for three crystal orientations. NI and NS are the spectral intensity for Si in the intermediate oxidation state and that for substrate silicon, respectively. The arrows show the existence of the data of (NI/NS) being smaller than the approximate value of the detection limit, 0.01. Oxide films were formed in dry oxygen.

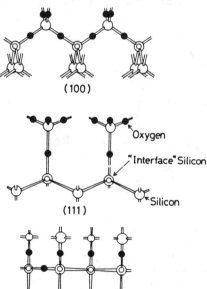

From the quantitative discussion on the extrapolated value, (NI/NS), obtained for zero dioxide film thickness in Fig. 2, it was shown[6] that Si-SiO_2 interfaces are abrupt for three crystal orientations discussed in the present paper. Possible interface structures are shown in Fig. 3.

In the discussion above, the interface structures are assumed to be independent of the oxide film thickness. This assumption is confirmed by the depth profiling measurements using chemical etching,[11] and also by measuring Si KLL Auger electron spectra.[11]

Fig. 3: Possible Si-SiO_2 interface structures for thermally grown oxide films.

Fig. 4 shows the effect of oxidation temperature on the thickness dependence of (NI/NS) for the oxide films formed on (100) surface in dry oxygen. According to this figure, the change in the oxidation temperature strongly affects the amounts of SiO and Si_2O_3 in the oxide film, while it weakly affects the amount of SiO at the interface.

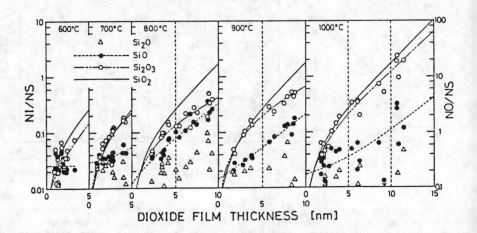

Fig. 4: Effect of oxidation temperature on thickness dependence of (NI/NS). Oxide films were formed in dry oxygen.

The change in oxidation atmosphere from dry oxygen to wet argon was found to diminish the amount of the Si_2O_3-like state in the oxide film. Also, the effect of oxidation temperature on the oxide films formed in wet argon was found to be different from that on the oxide films formed in dry oxygen.[12]

SUMMARY

Based on the observation of the crystal orientation dependence of Si 2p photoelectron spectra for ultrathin oxide films, the orientation dependence of the interface structures and the distribution of Si_2O_3 in the oxide film near the interface, but not at the interface. We conclude that Si-Si bonds are one of the possible structures for Si_2O_3. The effects of oxidation temperature and oxidation atmosphere on the amounts of intermediate oxides at and near the Si-SiO_2 interface were studied.

ACKNOWLEDGEMENTS

The authors are grateful to M. Ohno of Hitachi VLSI Engineering Co. Ltd. and to M. Ogino of Research and Development Center, Toshiba Corp. for supplying silicon wafers used in the present experiments. A part of this work was supported by the Grant-in-Aid for Special Project Research from the Ministry of Education, Science and Culture of Japan.

REFERENCES

1. For extensive references, see S. T. Pantelides, ed., The Physics of SiO_2 and Its Interfaces (Pergamon, New York, 1978).
2. R. Haight and L. C. Feldman, J. Appl. Phys. 53, 4884 (1982).
3. S. I. Raider and R. Flitsch, IBM J. Res. Dev. 22, 294 (1978).
4. G. Hollinger and F.J. Himpsel, Appl. Phys. Lett. 44, 93 (1984).
5. F. J. Grunthaner, P. J. Grunthaner, R. P. Vasquez, B. F. Lewis, J. Maserjian and A. Madhukar, Phys. Rev. Lett. 43, 1683 (1979).
6. T. Hattori and T. Suzuki, Appl. Phys. Lett. 43, 470 (1983).
7. N. M. Johnson, D. K. Biegelsen, M. D. Moyer, S. T. Chang, E. H. Poindexter and P.J. Caplan, Appl. Phys. Lett. 43, 563 (1983).
8. S. M. Goodnick, R. G. Gann, J. R. Sites, D. K. Ferry and C. W. Wilmsen, J. Vac. Sci. Technol. B1, 803 (1983).
9. T. Hattori, Y. Hisejima, H. Saito, T. Suzuki, H. Daimon, Y. Murata and M. Tsukada, Appl. Phys. Lett. 42, 244 (1983).
10. T. Suzuki and T. Hattori (unpublished).
11. T. Suzuki, M. Muto, M. Hara, T. Hattori, K. Yamabe and H. Yamauchi (unpublished).
12. T. Suzuki and T. Hattori (unpublished).

DYNAMIC CONDUCTIVITY MEASUREMENTS ON Si-SiO$_2$ INTERFACES DEGRADED BY OXIDE CHARGE AND AVALANCHE INJECTION

A. Zrenner and F. Koch
Physik-Department, Technische Universität München
D-8046 Garching, Fed. Rep. of Germany

EXTENDED ABSTRACT

In a paper to be published elsewhere /1/ we discuss the interface channel rf conductivity of MOS capacitor samples prepared on (100) p-Si wafers. We consider frequencies in the range 1-100 MHz. Such dynamic measurements are relevant for devices operated in a transient mode, with voltages switched and conductivities registered in a fraction of a μsec.

It has been demonstrated in ref. /2/ that at rf frequencies the surface conductivity can be measured in a contactless, capacitively-coupled fashion. For interfaces with locally inhomogeneous conductivity the rf measurement can be dominated by the capacitive shorting-out effect as in Fig. 1. There will be a distinct frequency-dependence of the impedance.

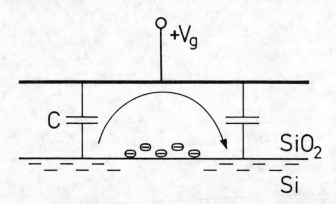

Fig. 1:
Capacitive by-pass of the rf. current at an interface region with trapped charge and vanishing conductivity. C represents the distributed oxide capacitance.

The paper considers two examples of inhomogeneous conductivity introduced by degradation of the interface. The first of these is the known interface potential fluctuation caused by the accumulation of randomly distributed positive charge ($\sim 10^{12}$ Na$^+$ ions/cm^2) at the Si-SiO$_2$ boundary. At low gate voltages conduction electrons assemble in disconnected islands in the regions of highest Na$^+$ concentration. We observe a steeply rising conductivity with increasing frequency as the conducting islands are connected by the capacitive by-pass.

The second case studied is that of an interface damaged by avalanche injection of electrons. A Na$^+$-drift with average density of only a few times 10^{10} cm^{-2} provides a locally fluctuating height of the interface barrier potential /3/. The injection of electrons is correspondingly inhomogeneous. As shown in Fig. 2 the electron conductivity at positive V_g has a marked frequency dependence. The hole conduction is insensitive to frequency. The observations are consistent with the creation of negatively charged interface defects capable of trapping a hole.

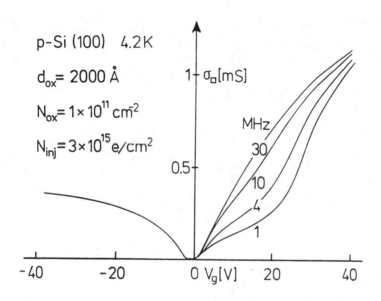

Fig. 2: Surface conductivity of electrons ($+V_g$) and holes ($-V_g$) after inhomogeneous avalanche injection

It is argued in ref. /1/ that because hot carrier effects in μm-sized device structures lead to local damage and channel inhomogeneity, the dynamic response can differ considerably from a conventional dc determination. If so, the operation of devices degraded by carrier injection in the locally strong fields near the drain would not be accurately predicted by the result of dc measurements.

/1/ A. Zrenner and F. Koch, J.Appl.Phys., submitted for publication
/2/ V. Dolgopolov, C. Mazuré, A. Zrenner, and F. Koch, J.Appl.Phys. **55**, 4280 (1984)
/3/ T.H. Di Stefano, Appl.Phys.Lett. **19**, 280 (1971)

THE ATOMIC STRUCTURE OF Si-SiO$_2$ INTERFACES SUGGESTING A LEDGE MECHANISM OF SILICON OXIDATION

J.H. Mazur and J. Washburn
Materials and Molecular Research Division
Lawrence Berkeley Laboratory, University of California, Berkeley, CA 94720

ABSTRACT

The atomic structure of Si-SiO$_2$ interfaces resulting from oxidation of singular $\{111\}$ and vicinal $(111)3°[1\bar{1}0]$ and $(111)2°[11\bar{2}]$ has been studied by high resolution electron microscopy. The transition from crystalline Si to amorphous SiO$_2$ was found to be very abrupt. The structure of the interface can be described by a terrace-ledge-kink-model. This structure is consistent with a ledge mechanism of silicon oxidation.

INTRODUCTION

The Si-SiO$_2$ system has been the subject of extensive investigations due to the application of thermally grown SiO$_2$ dielectric films in both bipolar and MOS technologies including large and recently very large scale integration.[1] Electronic processes at the Si-SiO$_2$ interfaces[2] which are important for MOS devices are known to be determined, in general, by interface structure which in turn is a function of the oxidation mechanism. This research attempts to establish an atomistic model of the Si-SiO$_2$ interface structure for different oxidizing conditions using high resolution electron microscopy (HREM) and from the interface structure deduce the oxidation mechanism.

EXPERIMENTAL PROCEDURES

Singular (111) and vicinal $(111)2°[11\bar{2}]$ and $(111)3°[1\bar{1}0]$ surfaces of CZ grown p-type, B-doped, 7-17 Ωcm silicon were oxidized at temperatures above 960°C in dry O$_2$. The wafers were cleaned before oxidation using procedures described elsewhere.[3] Before oxidation the native oxide was removed using 50:1 H$_2$O:HF solution followed by a rinse in deionized H$_2$O, and blow drying in N$_2$.

The wafers were immediately loaded in the oxidation furnace with argon or nitrogen flowing. After five minutes the ambient was changed to dry O$_2$ for the time necessary to grow about 100 nm of oxide. The wafers were then removed from the furnace again in the argon or nitrogen ambient within 2 minutes.

Cross-sectional transmission electron microscopy specimens were prepared by gluing two pieces of a wafer face to face with epoxy, and cutting such a sandwitch with a diamond saw normal to $\{110\}$. After grinding and double sided polishing to less than 100 μm, the section was glued to a support grid and ion milled to perforation at 5 kV and 15° incident angle. All observations were performed in a JEM 200 CX electron microscope equipped with a high resolution pole piece (C_s = 1.2 mm) operating at 200 kV. The specimens were imaged with the <110> Si substrate zone axis parallel to the electron beam. This configuration of the specimen allowed imaging of two sets of $\{111\}$ planes and one set of $\{200\}$ planes edge on.

EXPERIMENTAL RESULTS AND DISCUSSION

Oxidation of singular (the lowest surface energy) {111} Si surfaces at 1100°C in dry O_2 resulted in the Si-SiO$_2$ interface structure shown in Fig. 1. The oxide is amorphous as is revealed by a characteristic mottled contrast. The interface between silica and silicon is very abrupt and flat over the entire area observed except for the existence of ledges only one {111} interplanar distance (.314 nm) high. The ledges can be seen more clearly on the higher magnification micrograph shown in Fig. 2. The width of the terraces between positive and negative ledges varies and is dependent upon defocus which indicates that these ledges do not always extend through the entire TEM specimen thickness. Another interesting contrast feature that was observed is shown in Fig. 2. The last row of crystal image spots is displaced as would be expected if there was a stacking fault parallel to the surface. Computer modeling of the image is in progress in order to see if alternative explanations exist.

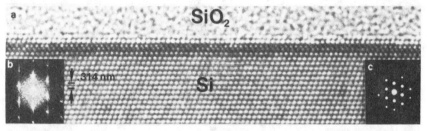

Fig. 1. (a) High resolution electron micrograph of a 100 nm thick oxide film grown on a singular (111) Si surface at 1100°C in dry O_2, (b) and (c) optical and selected area diffraction patterns respectively demonstrating imaging conditions.

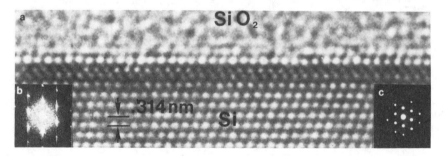

Fig. 2. Higher magnification image of the cross-section in Fig. 1 showing details of the ledge structrue at the interface and change of stacking order near Si-SiO$_2$ boundary, (b) optical diffraction pattern of the image (c) selected area electron diffraction pattern.

Figure 3 shows the Si-SiO$_2$ interface structure resulting from the oxidation at 1000°C of a vicinal (111)3°[1$\bar{1}$0] Si surface.[3] (Vicinal surfaces in contact with vacuum are expected to have ledges which connect terraces of minimum surface energy and the ledges would be expected to lie along low energy <110> directions). The transition from crystalline Si to amorphous SiO$_2$ in these specimens also takes place at flat (111) terraces about 6.0 nm wide, which are now separated by ledges all of the same sign, one interplanar distance

(.314 nm) high. This agrees very well with the calculated terrace width assuming ledges of one {111} interplanar spacing in height for a 3° inclination of the interface away from (111). A similar morphology at the Si-SiO$_2$ boundary has been observed for 2.0 nm thick native oxide formed on vicinal (111)3°[1$\bar{1}$0] Si surfaces at room temperature (Fig. 4).

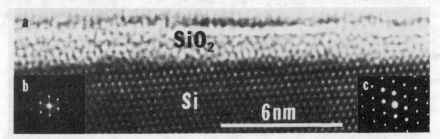

Fig. 3. High resolution image of a cross-section of a 100 nm thick oxide film grown on vicinal (111)3°[1$\bar{1}$0] Si surface in dry O$_2$, (b) optical diffraction pattern of the image (c) selected area electron diffraction pattern.

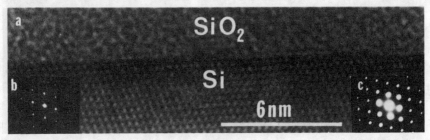

Fig. 4. (a) High resolution image of 2 nm thick native oxide on silicon, (b) optical diffraction pattern of the image, (c) selected area electron diffraction pattern.

The structure of the Si-SiO$_2$ interfaces resulting from oxidation at 1100°C of vicinal (111)2°[11$\bar{2}$] Si surfaces is shown in Fig. 5. In contrast to the more inclined Si surfaces some positive-negative ledge pairs on the terraces were observed. These step pairs are similar to those observed on singular {111} Si-SiO$_2$ interfaces.

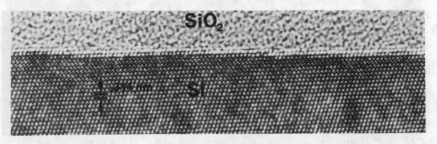

Fig. 5. High resolution image of a 100 nm thick oxide film grown on a vicinal (111)2°[11$\bar{2}$] Si surface.

The results of this investigation suggest that oxidation on {111} surfaces occurs layer by layer uniformly over large areas. Oxide growth apparently involves removal of Si atoms from the surface only at the ledges, probably at kinks. In the case of singular {111} surfaces, formation of the ledges must require repeated two dimensional nucleation corresponding to the formation of an oxide island in the next {111} layer of silicon. For vicinal surfaces, structural ledges are already present at the interface providing sites for oxidation. However for too low a density of such ledges two dimensional nucleation still was observed to take place resulting in terraces with additional positive and negative ledges. A similar process is observed for evaporation or dissolution of atoms from a surface into vapor or solution.[4,5] Although the Si surface in this case is in contact with solid silica, the interface structure appears to behave as it would in contact with a liquid. This is perhaps not surprising because viscous flow of silica occurs above 960°C[6] while oxidations in this work were performed at 1100 and 1000°C.

Thus a ledge mechanism for high temperature oxidation of silicon appears to be likely and is consistent with the observed interface structure.

CONCLUSIONS

These high resolution electron microscopy studies of the Si-SiO$_2$ interface structure resulting from oxidation of singular and vicinal (111) surfaces have demonstrated that:
(a) The Si substrate terminates abruptly on atomically flat (111) terraces at the Si-SiO$_2$ interface.
(b) The SiO$_2$ appears to be amorphous right up to the interface.
(c) Ledges one {111} interplanar distance (.314 nm) high are observed on both singular and vicinal oxidized surfaces.
(d) The observed structure suggest a terrace ledge-kink model for the interface and a ledge mechanism for high temperature oxidation similar to the mechanism of evaporation.

ACKNOWLEDGEMENTS

The authors acknowledge Dr. J. Plummer of Stanford University for providing the facilities for oxide growth and Ms. R. King for performing some of the oxidations. This work was supported by the Director, Office of Energy Research, Office of Basic Energy Sciences, Materials Science Division of the U.S. Department of Energy under Contract No. DE-AC03-76SF00098.

REFERENCES

1. S.M. Sze, Ed., VLSI Technology, McGraw Hill, N.Y. (1983).
2. S.M. Sze, Physics of Semiconductor Devices, 2nd ed., J. Wiley & Sons, N.Y. (1981).
3. J.H. Mazur, PhD Thesis, University of California, Berkeley, to be published.
4. D.K. Burton, et al., Phil. Trans. Roy. Soc. **243**, 299 (1950).
5. J.P. Hirth in Energetics in Metallurgical Phenomena, vol. 2 W.M. Mueller, Ed., Gordon and Breach, N.Y. (1965), p. 1.
6. E.P. EerNisse, Appl. Phys. Lett., **35**, 8, (1979).

Part III. Interconnections and Silicides

LASERS FOR INTERCONNECTIONS, CIRCUIT FABRICATION AND REPAIR

R. J. von Gutfeld
IBM T. J. Watson Research Center, Yorktown Heights, NY 10598

ABSTRACT

A review of lasers and their applications in processing microelectronic materials on both the wafer and packaging level is presented. Schemes for wiring and implemeting redundancy on VLSI wafers using highly focused laser pulses are discussed. A variety of maskless deposition and etching techniques together with their physical mechanisms are reviewed, including those relating to the deposition of gold, copper and silicon as well as the localized removal of metals, semiconductors and polymers.

INTRODUCTION

One of the very useful features of lasers is the small spot size to which these beams can be focused, i.e., on the order of a wavelength of light. The diffraction limited spot size in combination with high intensity nanosecond pulses make it possible to alter microcircuits on the chip level. Similarly, larger spot sizes in conjunction with longer laser pulse durations or even cw beams can be utilized to alter certain less critical regions on a wafer as well as circuit boards, i.e., the packaging associated with these VLSI chips. In certain cases, the interactions of the laser with gases and liquids can actually be a part of the VLSI manufacturing process. The effects to be discussed in this review include the use of the laser as a heat source to cause localized melting and ablation of circuit elements for connection and disconnection purposes. In addition, a variety of techniques will be described that use laser energy to bring about pyrolytic, photothermal and photochemical effects on a very localized, i.e. submillimeter to micrometer scale. With these techniques, maskless patterning is achieved both by the deposition and removal of material. A large variety of materials have been affected using these laser processing techniques including the deposition of nickel, aluminum, gold, copper, silicon, SiO_2, chromium, iron, tungsten, etc. Localized removal of Cu, Al, Si, GaAs and a variety of polymers has also been achieved. Alloying by simultaneous deposition and melting has been reported for the purpose of making ohmic contacts and forming p-n junctions.

The intent of this report is to present brief discussions of a number of the aforementioned techniques and processes with examples of how they may be applied to VLSI. The applications will pertain to both chip and associated VLSI packaging such as boards and connectors.

More detailed discussions of this subject material can be found both in the referenced papers and in a number of conference proceedings devoted in part to laser processing of microelectronic materials.[1-5] In addition, this work has in part been summarized in two book chapters which contain numerous additional references.[6,7]

0094-243X/84/1220056-07 $3.00 Copyright 1984 American Institute of Physics

SILICON CIRCUIT ALTERATIONS: LASER MELTING AND ABLATION OF METAL LINES

Two types of circuit changes have been successfully achieved using short laser pulses applied directly to logic and memory circuits. Disconnects or line opening has been accomplished by using nanosecond pulses from a Q switched Nd-Yag laser to vaporize aluminum metal lines to wire the desired memory configuration on read-only Si on sapphire memories.[8] Somewhat later, initial studies on MOS structures were made with 3-5 ns nitrogen laser pumped dye laser pulses (rhodamine 6G) with power densities up to $2 \times 10^9 \text{W/cm}^2$.[9,10] Both connection and disconnections were made on bipolar transistor circuits. In the connection process two or more focused laser pulses were directed onto an aluminum line, separated by approximately a 1 micron layer of SiO_2 from an underlying n$^+$ diffusion layer. These initial pulses served to vaporize the line locally, thereby exposing the SiO_2 layer. A subsequent pulse or series of pulses caused the optically absorbing Si to melt and explosively rupture the insulating layer. A solid fillet of Si results from rapid cooling, probably alloying with the Al to form an ohmic contact. A similar line connection concept has been employed to connect metal-to-metal lines separated by an SiO_2 insulation layer. Aluminum to Al laser connections were studied by Logue et al[11] for optimizing engineering designs of logic arrays using a basic chip called a laser programmable logic array (LPLA). Using these laser connections at prescribed points this group was able to develop and test new circuit designs without resorting to mask changes or new chip fabrication after each trial. The metal-to-metal connections were made much in the same way as just described for the bipolar circuits except that here the Al made ohmic connection to the overlying Al line through the insulator. This laser personalization process provided rapid turn-around time and showed for the first time the utility of the laser for MOS VLSI design studies.

A much longer pulse width, 1 ms, has been used by Raffel et al on a CMOS integrator to connect metal-to-metal (aluminum) through a thin oxide - α - silicon - oxide insulator using a focused argon laser of approximately 1 watt.[12] The cutting of lines on the same integrator has also been demonstrated. It should be pointed out that here the points of connection and disconnection were far removed from the transistor circuits so that even with these relatively long pulses no thermal damage to these circuits resulted. Linking of two metal lines laterally, i.e. two lines separated by a polyimide gap has also been demonstrated. Here, the laser pulses cause the polyimide to become conducting, thereby connecting the two separated lines.[13]

The use of laser cutting of lines for discretionary wiring of read-only memories has been demonstrated by North and Weick while similar techniques have been used by a number of groups to implement redundancy of random access memories (RAM), mostly in conjunction with Q-switched Nd-YaG lasers.[14-17] Redundancy implementation has as its main purpose wafer yield improvement particularly in the early stages of manufacture. The repair of malfunctioning memories is brought about by having spare parts on the wafer which can be connected by a simple laser delete process to replace the defective element. The defective element can also be made inoperative by appropriate laser delete steps. Very high yield improvements have been demonstrated utilizing this technique as reported in some detail in Ref. 17.

A pulsed XeCl pulsed excimer laser has been used by Andrew et al[18] for the patterning of metal films over large areas on various substrates. The patterning is accomplished with a single pulse (10 ns) of this 308 nm light at relatively low fluences of 0.05 -0.24 J/cm^2. Silver, gold, aluminum, nickel, copper and chromium have been successfully removed in this manner by a mechanism believed to be increased pressure at the film-substrate interface with film temperatures below that of film vaporization. Large area patterning can be achieved with the use of projection optics.

LASER DEPOSITION AND ETCHING - GASEOUS ATMOSPHERES

Lasers interact in a number of ways with gases and gas-substrate interfaces which can give rise either to deposition of gas atoms onto the surface or removal, i.e., etching of surface atoms by a variety of mechanisms. Resonant excitation of a gas molecule via multiphoton absorption can give rise to highly excited vibronic levels which lead to dissociation. Dissociation can also occur from dielectric breakdown of the gas resulting from high intensity fluxes. Both of these mechanisms can give rise to deposition or etching depending on the gases and particular details of the system. Local surface heating by the laser can lead to pyrolytic dissociation of the gas molecule and result in laser chemical vapor deposition (LCVD). A particularly important advantage of LCVD over ordinary CVD is that only very small areas become heated which can greatly enhance the purity of the deposition. Radiation can also enhance nucleation sites and increase sticking coefficients of depositing atoms. Etching can result from the formation of excited gas atoms, which for certain gases become highly reactive. These atoms interact with the surface to produce a volatile compound.

The technique of depositing or removing atoms or molecules from a surface in a gas atmosphere has been termed "direct writing" since the laser beam determines where the local microreaction will take place so that patterning occurs without a mask. Numerous examples of laser "direct write" depositions are cited in the literature. Elements such as Fe, W, and Cr have been deposited from their respective carbonyls using UV and IR lasers. Lines as small as $2.5\mu m$ wide have been produced by moving the sample (within a chamber containing the gas atmosphere) with respect to the focused laser beam.[19] Metal-alkyl compounds have also served as a source of metal atoms that have been laser deposited such as aluminum and cadmium, derived from the gases trimethylaluminum and dimethylcadmium respectively. Both the 193 nm excimer laser and a frequency doubled argon laser have been used as laser sources for these depositions.[20,21]

The chemical vapor deposition of poly-crystalline and amorphous silicon from silane using both pulsed and cw CO_2 lasers has been reported by several groups.[22-24] In these experiments the laser beam was directed onto the substrate to cause heating or aimed parallel and slightly above the substrate to produce vibronic excitation of the gas with an additional substrate heater to promote LCVD. Doped silicon films have been produced from a mixture of silane containing either BCl_3 or $B(CH_3)_3$. The resulting films are heavily boron doped poly-silicon and good electrical conductors. Film dimensions as small as $1\mu m$ widths have been achieved from thermal decomposition with the use of a focused argon laser.[25] Amorphous silicon grown by LCVD is relatively defect free due to a relatively high inclusion of hydrogen which serves to tie up the dangling bonds. This higher hydrogen content occurs in part due to the fact the LCVD silicon can be grown at lower temperatures than CVD silicon. This allows LCVD a-silicon to be doped for use in large area solar cells and thin film transistors.[23,24] Other materials deposited by LCVD include Ni, TiO_2 and TiC.[26] Needles and thin films of carbon have been grown from the pyrolytic dissociation of C_2H_2.[27] Recently, gold films have been deposited by LCVD derived from metal bearing organic molecules.[28]

Additional applications for the aforementioned gas depositions relate to 1) mask repair, 2) circuit repair for both wafers and VLSI packaging, 3) the formation of p-n juncitons via simultaneous deposition and liquid phase diffusion for use with GaAS solar cells [29], 4) high conducting poly-silicon lines that have been used for interconnects on MOS structures,[30] 6) metal-oxide-semiconductor capacitors, fabricated by reacting SiH_4 and N_2O gas molecules to form SiO_2 via photochemical deposition.[31]

Extensive work on the etching of Si, SiO_2, ceramics and numerous metals has been reported over the last several years. Many of these results are summarized in Ref. 1-7, and will not be enumerated here in detail. An interesting application utilizing both laser gaseous deposition and etching techniques is that of fabricating n-MOS transistors. Termed "wafer-scale pantography", this method is an attempt to fabricate an entire wafer utilizing only laser processes in conjunction with computer controlled process operations.[32]

ETCHING IN AIR

Considerable interest has been found in the use of excimer (pulsed UV) lasers and their application for the dry etching of numerous polymers.[33,34] It is belived that in the far ultra violet, that is, at wavelenghts below ~ 200nm, chemical bond breaking is the predominant mechanism which accounts both for the material removal rate dependence on incident laser fluence and the observed fluence threshold prior to substantial removal. However, at longer wavelengths, for example 308 nm, a thermal model has been proposed to explain the observations for the same materials as in Ref. 33.[35] A number of polymers have been patterned by directing excimer laser light through a mask. These include polyethylene terephthalate (PET), kapton, polyimide[36], nitrocellulose and polymethylmethacralate.[37,38] For PET, etch rates of 1200Å/ pulse have been reported for the 193 nm laser (pulse width ~ 14ns with an incident fluence of $370 mJ/cm^2$[33]). A pattern with $0.3\mu m$ resolution has been produced in nitrocellulose for the same wavelength of light.[38] Here, a 2000Å/ pulse removal rate was measured for a fluence of only 50 mJ/pulse. [38] The importance of this single step technique for dry processing of photoresist materials continues to be a subject of very active investigation.

PHOTOTHERMAL EFFECTS IN LIQUIDS - LASER ENHANCED PLATING AND ETCHING

A number of metals have been deposited locally from solution at very high rates by shining a focused laser onto a cathode in a plating cell.[39] This maskless patterning technique is, in many ways, analogous to the "direct write" technique for gaseous systems except that for laser enhanced electroplating a negative voltage is applied between the cathode and the counter electrode (anode). Those parts of the cathode irradiated by the focused laser beam have shown plating rates 3 to 4 orders of magnitude greater than the non-illuminated regions, that is, the background plating rate. The three elements plated from solution and studied in some detail are copper, nickel and gold. While copper and nickel can be derived from simple salt solutions, gold is generally complexed to a cyanide radical. The gold solutions were obtained from commercial suppliers. Copper lines as small as $2\mu m$ in width and copper plating rates up to $10\mu m/s$ have been reported. The photothermal plating mechanisms have been investigated in some detail and identified as: 1) a shift in the electrochemical rest potential with increasing temperature produced by the laser absorption at the cathode. This shift is toward a more positive value for copper and gold and increases the plating rate; 2) Increased charge transfer rates with increasing temperature, and 3) increased mass transport due to strong thermal gradients which cause local microstirring leading to a replenishment of ions in the diffusion layer.[40]

It has also been possible to obtain maskless depositions without the use of externally applied potentials (electrodeless plating). Electroless plating is a particular example of electrodeless plating and these solutions contain a catalyst which maintains charge neutrality and allows the deposition to occur without an external driving source.[40,41] Laser enhanced exchange plating on the other hand also permits high speed patterning without external potentials but here the local shift in the rest potential causes plating to occur in the hotter region (laser irradiated) while simultaneous etching occurs elsewhere on the cooler metal surface. Plating of both copper and gold has been reported using this technique. [40,41]

Recently, laser enhanced jet plating has been reported as a means for increasing the rate of deposition over and beyond that achieved by laser enhanced plating.[42] This maskless patterning scheme utilizes a pressurized free standing stream of electrolyte directed through a nozzle at the cathode. A laser beam is focused into this stream collinearly so that both the light and fluid are incident on the cathode at the same point. Application of a voltage between the nozzle (anode) and the cathode causes current flow to occur through the stream. The jet thereby serves as a source of rapid ion resupply, a means for producing localized maskless plating due to the localization of current flow at the cathode defined by the cross-sectional area of the jet, and as an optical waveguide which confines the light beam. It has been found that the laser enhanced jet can deposit gold at rates up to $20\mu m/s$ on microelectronic connector materials such as Be-Cu to produce gold contact areas on the order of $\sim 1/2$ mm in diameter. This deposition rate is more than an order of magnitude faster than the fastest non-laser jet plating rates for quality gold depositions and over two orders of magnitude faster than the fastest present day conventional gold plating techniques for gold. Potentially large gold savings can be realized with this deposition method since standard rack plating is ineffective in producing localized plating and gives rise instead to large areas of unnecessary precious metal deposition.

Applications for laser enhanced plating in addition to those mentioned above are generally in the areas of packaging for circuit design and repair. Deposition rates are generally considerably faster than those obtained from photolytic laser direct-write techniques. The plating techniques are especially useful since they offer a means for depositing two very important metals for VLSI applications at very high speeds, copper and gold.

Laser etching in liquids has been studied using both electrochemical and chemical techniques. Ehrlich, et al. have demonstrated laser enhanced chemical etching of Al using 120mW of focused argon laser light in combination with a nitric acid mixture to alter gate arrays on CMOS structures. A spatial resolution of better than $2\mu m$ has been achieved.[30,43]

REFERENCES

1. Technical Digest, Conference on Lasers and Electrooptics (CLEO 1982) April 13-15, 1982, Phoenix, AZ
2. Abstracts, Materials Research Society, Nov. 1-4, 1982, Boston, MA
3. Technical Digest, Conference on Lasers and Electrooptics (CLEO 1983) May 17-20, 1983, Baltimore, MD
4. Abstracts Materials Research Society, Nov. 14-17, 1983, Boston, MA
5. Technical Digest, Conference on Lasers and Electrooptics (CLEO 1984) June 19-22, 1984, Anaheim, CA
6. R. J. von Gutfeld in "Laser Applications", Vol. 5, J. F. Ready and R. K. Erf, Eds. Chapter I, pp. 1-67, Academic Press (1984).
7. D. J. Ehrlich and J. Y. Tsao in VLSI Electronics: Microstructure Science Vol. 7, N. Einspruch, Ed. Academic Press (1984).
8. A. D. Sypherd and N. D. Salman, Proc. Natl. Electron. Conference, **24**, 206-208 (1968).
9. P. W. Cook S. E. Schuster and R. J. von Gutfeld, Appl. Phys. Lett. **26**, 124 (1975).
10. L. Kuhn, S. E. Schuster, P. S. Zory, G. W. Lynch and J. T. Parrish, IEEE J. Solid-State Circuits, **SC-10**, 219 (1975).
11. J. C. Logue, W. J. Kleinfelder, P. Lowry, J. R. Moulic and W. W. Wu, IBM J. of Res. Dev. **25**, 107 (1981).
12. G. H. Chapman, A. H. Anderson, K. H. Konkle, B. Mathur, J. I. Raffel and A. M. Soares, Tech. Digest CLEO Conf. Proc., p. 222, 1984 Anaheim, CA.
13. J. I. Raffel, J. F. Freidin and G. H. Chapman, Appl. Phys. Lett. **42**, 705 (1983).
14. J. C. North and W. W. Weick, IEEE J. Solid-State Circuit **SC-11** p. 500, (1976).
15. J. F. M. Bindels, J. D. Chlipala, F. H. Fischer, T. F. Mantz, R. G. Nelson and R. T. Smith, Dig. Tech. Pop. Int. Solid-State Circuit Conf., p. 82 (1981).
16. R. P. Cenker, D. G. Clemons, W. R. Huber, J. B. Petrizzi, F. J. Procyk and G. M. Trout, Dig. Tech. Pop. Int. Solid-State Circuit Conf., p. 260 (1979).
17. R. T. Smith, J. D. Chlipala, J. F. M. Bindels, R. G. Nelson, F. H. Fischer and T. F. Mantz, IEEE J. Solid State Circuits **SC-16**, 506 (1981).
18. J. E. Andrew, P. E. Dyer, R. D. Greenough and P. H. Key, Appl. Phys. Lett. **43**, 1076 (1983).
19. D. J. Ehrlich, R. M. Osgood, Jr. and T. F. Deutsch, J. Electrochem. Soc. **128**, 2039-2041 (1981).
20. T. F. Deutsch, D. J. Ehrlich and R. M. Osgood, Jr., Appl. Phys. Lett. **35**, 195 (1979).
21. D. J. Ehrlich, R. M. Osgood, Jr. and T. F. Deutsch, Appl. Phys. Lett. **36**, 916 (1980).
22. C. P. Christensen and K. M. Lakin, Appl. Phys. Lett. **32**, 254 (1978).

23. M. Hanabusa, A. Namiki and K. Yoshihara, Appl. Phys. Lett. **35**, 626 (1979); also M. Hanabusa, H. Kikuchi and T. Iwanaga, CLEO Conf. Proc. p. 156, 1984 Anaheim, CA.

24. M. Meunier, T. R. Gattuso, D. Adler and J. S. Haggerty, Appl. Phys. Lett. **43**, 273 (1983).

25. D. J. Ehrlich, R. M. Osgood, Jr., and T. F. Deutsch, Appl. Phys. Lett. **39**, 957 (1981).

26. S. D. Allen, J. of Appl. Phys. **52**, 6501 (1981).

27. G. Leyendecker, D. Bauerle, P. Geittner and H. Lydtin, Appl. Phys. Lett. **39**, 921 (1981).

28. F. A. Houle, C. R. Jones, T. H. Baum and C. A. Kovac, CLEO Conf. Proc., p. 156 (1984) Anaheim, CA.

29. T. F. Deutsch, J. C. C. Fan, D. J. Ehrlich, G. W. Turner, R. L. Chapman and R. P. Gale, Appl. Phys. Lett. **40**, 722 (1982).

30. D. J. Ehrlich, J. Y. Tsao, D. J. Silversmith, J. H. C. Sedlacek, R. W. Mountain and W. S. Graber, IEEE Electron. Dev. Lett., **EDL-5**, 32 (1984).

31. P. K. Boyer, G. A. Roche, W. H. Ritchie, and G. J. Collins, Appl. Phys. Lett. **40**, 716 (1982).

32. B. M. McWilliams, I. P. Herman, F. Mitlitsky, R. A. Hyde and L. L. Wood, Appl. Phys. Lett. **43**, 946 (1983).

33. R. Srinivasan and V. Mayne-Banton, Appl. Phys. Lett. **41**, 576 (1982).

34. B. J. Garrison and R. Srinivasan, Appl. Phys. Lett. **44**, 849 (1984).

35. J. E. Andrew, P. E. Dyer, D. Forster and P. H. Key, Appl. Phys. Lett. **43**, 717 (1983).

36. P. E. Dyer and J. Sidhu, CLEO Conf. Proc. p. 188 (1984) Anaheim, CA.

37. M. W. Geis, J. N. Randall, T. F. Deutsch, P. D. DeGraff, K. E. Krohn, and L. A. Stern, Appl. Phys. Lett. **43**, 74 (1983).

38. T. F. Deutsch and M. W. Geiss, J. Appl. Phys. **54**, 7201 (1983).

39. R. J. von Gutfeld, E. E. Tynan, R. L. Melcher and S. E. Blum, Appl. Phys. Lett. **35**, 651 (1979); also R. J. von Gutfeld, E. E. Tynan and L. T. Romankiw, Meeting Electrochem Soc. Extended Abstract No. 472, p. 1185 (1979).

40. J. Cl. Puippe, R. E. Acosta and R. J. von Gutfeld, Journ. Electrochem. Soc. **128**, 2539 (1981).

41. R. J. von Gutfeld, R. E. Acosta and L. T. Romankiw, IBM Journ. Res. and Dev. **26**, 136 (1982).

42. R. J. von Gutfeld, M. H. Gelchinski, L. T. Romankiw and D. R. Vigliotti, Appl. Phys. Lett. **43**, 876 (1983).

43. J. Y. Tsao and D. J. Ehrlich, Appl. Phys. Lett. **43**, 146 (1983).

A STUDY OF LOW TEMPERATURE Pd AND Ni SILICIDES
FORMED ON DRY ETCHED SILICON SURFACES

X. C. Mu, A. Climent,* and S. J. Fonash
Engineering Science Program
The Pennsylvania State University
University Park, PA 16802

ABSTRACT

As-deposited palladium and nickel contacts to dry etched (reactive ion etched or ion beam etched) silicon surfaces display anomalous current-voltage (I-V) behavior due to the damage layer induced by the dry etching processing. This study examines the possibility that normal current-voltage behavior may be recovered by consuming the damage during low temperature silicide formation. Evolution of Pd/Si and Ni/Si contacts for different annealing times was monitored by means of current-voltage, capacitance-voltage, and low temperature (<0°C) activation energy. It is found that damage still remains for both Pd/Si and Ni/Si systems after low temperature silicide formation; however, this damage is more noticeable on n-type than it is on p-type silicon.

INTRODUCTION

Work in our group using Rutherford backscattering spectroscopy (RBS) and electron spin resonance (ESR) has established that dry etching (reactive ion etching (RIE) or ion beam etching (IBE)) of silicon produces a modified layer containing lattice damage. Further, it has been demonstrated by our group and others that metal contacts (using non-reactive metals) to these layers and MOS structures fabricated on these layers display anomalous electrical behavior[1-4]. Current approaches used in production for coping with this dry etching damage problem include removing the damaged layer with a final wet etch (clearly unacceptable for fine line structures) and high temperature furnace annealing (becoming more unacceptable with the drive to lower temperature processing). We have also demonstrated that the damage in these dry etched layers can be passivated by low energy hydrogen ion implantation.[5]

The study reported here focuses on determining if low temperature (200 ∿ 300°C) silicide formation can consume the modified layer produced by dry etching. The 200 ∿ 300°C temperature range employed is ideally compatible with the drive to low temperature processing. Its use in our study has the further advantage that dry etching damage consumption in silicide formation can be separated from damage removal by annealing (damage annealing does not occur at these temperatures). Contact changes during silicide formation on these dry etched surfaces were monitored by following the activation energy

* Visiting Fulbright Scholar from Universidad Autonoma de Madrid, Canto Blanco, Madrid 34, Spain.

(for temperatures <0°C) for the current-voltage (I-V) characteristics. In some cases the Schottky barrier height evolution of these Pd/Si and Ni/Si contacts for different annealing times was also monitored by means of the capacitance-voltage (i.e., $1/C^2$ vs. V) technique.

EXPERIMENTAL PROCEDURE

All the samples subjected to IBE were either <111> or <100> Monsanto, Czochrolski-grown silicon. Both n- and p-type silicon were used with resistivities between 1 and 10 Ω-cm. The IBE process was done using a Kaufman ion source in a Commonwealth Scientific ion-beam etching unit. The chamber was pumped down to $\sim 2 \times 10^{-6}$ torr and then backfilled with the etching species (Ar) to a pressure of $\sim 4 \times 10^{-4}$ torr. The system is then purged for 10 minutes. The etching time employed was always kept to 5 minutes and ion beam energy was 0.4 KeV. In all cases, the extractor voltage was 300 V, the accelerator current 200 mA and the glow current 4 to 5 A. The substrate holder was water-cooled and was rotated during etching in order to give uniformity.

All samples subjected to RIE were <111> Monsanto, Czochralski-grown p-type silicon with resistivity of $4 \sim 8$ Ω-cm. The RIE processing was done in an Anelva 503 reactor using CCl_4 at a total pressure of 20 m torr and an RF power of 600 W.

Immediately prior to metal deposition, the etched samples (RIE and IBE) together with controls (i.e., samples only chemically etched but not dry etched) were given a mild chemical etch in diluted ($\sim 5\%$) HF, rinsed in deionized water, and blown dry with N_2 to remove any native oxide. The Pd and Ni depositions were performed in an ionization-pumped vacuum system. The base pressure during evaporation was maintained in the low 10^{-7} torr range. An aluminum deposition on the back surfaces was used for ohmic contacts.

Annealing treatments for palladium and nickel silicide formation were performed in an oil diffusion-pumped vacuum system employing a liquid nitrogen trap to avoid backstreaming. The pressure during the annealing was kept in the mid 10^{-6} torr range. The annealing temperatures were between 240°C and 260°C and the annealing times were between 10 and 30 minutes.

I-V measurements were performed on all samples. C-V barrier height measurements were only done on n-type IBE samples since the conductance of palladium and nickel contacts to p-type silicon was too high to get meaningful capacitance values. Low temperature (<0°C) activation energy measurements were performed on all IBE samples.

RESULTS AND DISCUSSION

A. <u>I-V Characteristics</u> are always a very sensitive probe to study the Schottky barrier height ϕ_B. The I-V characteristics of palladium and nickel contacts on silicon subjected to RIE or IBE were examined prior to any annealing steps and were re-examined after various annealing times to evaluate the effect on ϕ_B due to silicide formation. Before annealing, the I-V characteristics of Pd/Si and Ni/Si contacts on dry etched surfaces show barrier lowering for n-type

Si, when compared to controls on only chemically etched n-type Si, and showed barrier height increasing for p-type Si, when compared to controls on only chemically etched p-type Si. In Figures 1 through 6, we show these results denoted as 0 min. annealing (i.e., without any annealing treatment). These results indicate that positive charge is induced in the modified layer for both Pd and Ni/dry-etched-Si systems. The same effect has been found to be present for Au/dry-etched-Si system (1, 2).

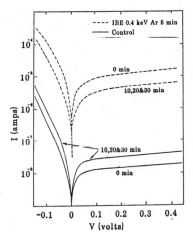

Fig. 1. I-V characteristics of Pd/n-Si<100> contacts for different anneal times at 260°C (all 1 mm dots).

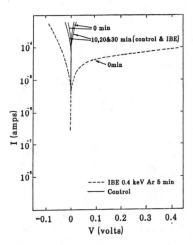

Fig. 2. I-V characteristics of Pd/p-Si<100> contacts for different anneal times at 260°C (all 1 mm dots).

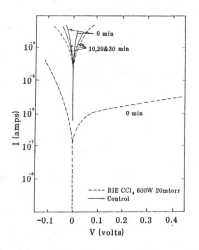

Fig. 3. I-V characteristics of Pd/p-Si<111> contacts for different anneal times at 260°C (all 1 mm dots).

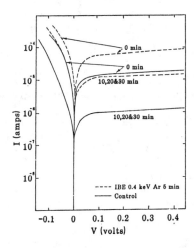

Fig. 4. I-V characteristics of Ni/n-Si<100> contacts for different anneal times at 260°C (all 1 mm dots).

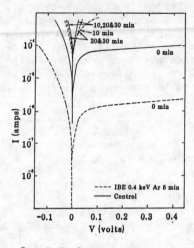
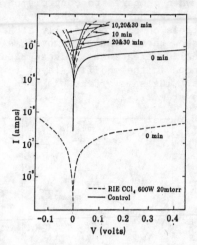

Fig. 5. I-V characteristics of Ni/p-Si<100> contacts for different anneal times at 260°C (all 1 mm dots).

Fig. 6. I-V characteristics of Ni/p-Si<111> contacts for different anneal times at 260°C (all 1 mm dots).

After heat treatment at about 260°C for various times (10, 20, and 30 min.), the I-V characteristics of these samples showed different behavior. For Pd/Si system, one may deduce from Figs. 1, 2, and 3 that the Schottky barrier height <u>decreases</u> on n-type Si but <u>increases</u> on p-type Si with silicide formation for chemically etched Si. One the other hand, it may be deduced that the Schottky barrier height <u>increases</u> on n-type Si but <u>decreases</u> on p-type Si with silicide formation for dry etched Si (see Figures 1, 2, 3). For Ni/Si system, barrier height <u>increases</u> on n-type Si but <u>decreases</u> on p-type Si for <u>both</u> chemically etched and dry etched Si (see Figures 4 to 6).

We note that the barrier heights of both Pd/Si and Ni/Si systems recover well on dry etched <u>p-type</u> Si using low temperature silicide growth. However, they do not recover fully to the expected behavior on <u>n-type</u> Si using low temperature silicide growth.

B. <u>C-V Measurements</u> were done on n-type Si in order to directly measure changes in the Schottky barrier height rather than inferring them from I-V data. Only IBE samples of 0.4 keV Ar etch and controls for Pd/Si and Ni/Si were used. As discussed earlier, n-type Si was used for this measurement. The trend of barrier height recovery as a function of annealing time was found to agree very well with that inferred from the I-V characteristics.

C. Low temperature (<0°C) <u>activation energy measurements</u> of both forward and reverse bias transport (at + 0.1 V and −0.5 V respectively) were performed on all controls and IBE samples. We present the results in Figs. 7 to Fig. 10. Again the Schottky barrier height recovery trend agrees with that from I-V and C-V measurements. Note that for the Pd/n-Si system, there is a significant difference in the activation energy for forward-bias and reverse-bias. This is probably due to the fact that Schottky barrier of Pd/n-Si is relatively high so that the reverse leakage current is not dominated by the thermionic emission current component but by a leakage current component. The situation in Pd/p-Si and Ni/Si system is seen to be much better.

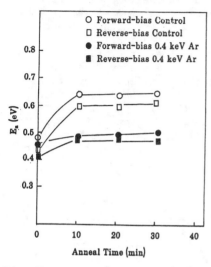

Fig. 7. Low temperature (<0°C) activation energy measurements of Ni/n-Si<100> for different anneal times at 260°C (all 1 mm dots).

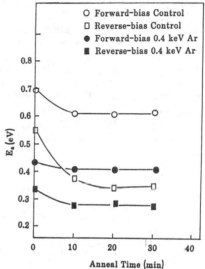

Fig. 8. Low temperature (<0°C) activation energy measurements of Pd/n-Si<100> for different anneal times at 260°C (all 1 mm dots).

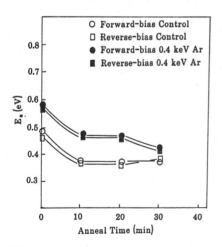

Fig. 9. Low temperature (<0°C) activation energy measurements of Ni/p-Si<100> for different anneal times at 260°C (all 1 mm dots).

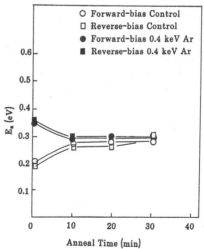

Fig. 10. Low temperature (<0°C) activation energy measurements of Pd/p-Si<100> for different anneal times at 260°C (all 1 mm dots).

CONCLUSIONS

Palladium and nickel contacts to dry etched (RIE and IBE) silicon surfaces display anomalous I-V behavior. This anomalous I-V behavior recovers well to the expected behavior with low temperature silicide formation for p-type silicon while it does not fully recover with silicide formation at low temperatures for n-type silicon. This result indicates that the damage induced during dry etching process still remains after low temperature palladium and nickel silicide formation. By examining the Schottky barrier height evolution for various annealing times, we noted that for Pd/Si system, the Schottky barrier height decreases on n-type silicon with silicide formation for chemically etched silicon and increases on n-type silicon with silicide formation for dry etched (RIE or IBE) silicon. The inverse holds for p-type silicon. For Ni/Si system, the Schottky barrier height increases on n-type silicon with silicide formation for both chemically etched and dry etched silicon. Again the inverse is true for p-type silicon.

REFERENCES

1. S. J. Fonash, S. Ashok, and R. Singh, Thin Solid Films, 90, 231 (1982).
2. R. Singh, S. J. Fonash, S. Ashok, P. J. Caplan, J. Shappirio, M. Hage-Ali, J. P. Ponpon, J. Vac. Sci. Technol., Vol. A1, p. 334 (1983).
3. S. Ashok, T. P. Chow, and B. J. Baliga, Appl. Phys. Lett., Vol. 42, p. 687 (1983).
4. S. W. Pang, D. D. Rathman, D. J. Silversmith, R. W. Mountain, and P. D. DeGraff, J. Appl. Phys., Vol. 54, p. 3272 (1983).
5. R. Singh, S. J. Fonash, P. J. Caplan, and E. H. Poindexter, Appl. Phys. Lett., Vol 43, p. 502 (1983).
6. N. Cheung, S. S. Lau, M. A. Nicolet, J. W. Mayer, and T. T. Sheng, Proc. Symp. on Thin Films, Interfaces and Interactions, Vol. 30-2, Electro. Chem. Soc.
7. A. Climent and S. J. Fonash, J. of Appl. Phys., Vol. 56(4), 1063-1069 (1984).

Part IV. Lithography

LITHOGRAPHY REQUIREMENTS IN COMPLEX VLSI DEVICE FABRICATION

Alan D. Wilson

IBM Thomas J. Watson Research Center
Yorktown Heights, NY 10598

ABSTRACT

Fabrication of complex VLSI circuits requires continual advances in lithography to satisfy: decreasing minimum linewidths, larger chip sizes, tighter linewidth and overlay control, increasing topography to linewidth ratios, higher yield demands, increased throughput, harsher device processing, lower lithography cost and a larger part number set with quick turn around time.

I will discuss where optical, electron beam, x-ray and ion beam lithography can be applied to judiciously satisfy the complex VLSI circuit fabrication requirements and address those areas that are in need of major further advances. Emphasis will be placed on advanced electron beam and storage ring x-ray lithography.

INTRODUCTION

In the past 25 years, the field of integrated semiconductor circuits has grown in complexity approximately six orders in magnitude and has become a multi-billion dollar business that is as pervasive as the automobile. This phenomenal complexity growth has occurred, in part, because of our constant reduction in the size of minimum features used to fabricate individual transistors and memory cells. We find that about each five to six years the resist image minimum feature is halved and the chip area doubles. Each three years a new generation of memory chips with a 4x increase in bit size emerges and each six years the birth of two memory generations with a 16x increase in memory bits occurs. By early 1990's, production devices may have $0.5\mu m$ features and be assembled into circuits on 100 - 200 mm^2 chips. The functional complexity of 1990 chips are hence expected to be 8 - 10x of those introduced next year (1985) into manufacturing. Sixteen to over 64 Mbit memory chips do not seem unreasonable from the point of view of the lithography alone (ignoring some serious defect questions). It can also be argued that it has been the constant demand for increased circuit complexity that has driven over the past quarter century the minimum feature sizes from "mils" to about a micron. You may justifiably ask why all this emphasis on smallness - the answer is: cost and performance. A question many of us argue, ponder and frequently try to answer is "will it ever end?" and perhaps just as frequently, "will optics make it?"

In this brief article we will again debate the issues of submicron ($.9 \rightarrow .1\mu m$) lithography applied to the fabrication of complex chips in manufacturing near the end of the present decade and into the next decade and if our progress is slowed, perhaps beyond year 2000 (assuming we do not get to sub 100 nanometer ($.1\ \mu m$) until after 2000). This debate covers limiting aspects of competing lithography techniques (optical, electron-beam, x-ray and ion beam) for submicron feature size definition. The discussion is restricted to "direct" lithographic techniques with the exclusion of non-direct methods such as thin films (sidewalls), shadowing, etc. methods for fabrication of selective features.

Lithography is a diffuse subject that eludes some, mistifies many and presents captivating problems to a few of us. Lithography is a subject which requires a "systems" approach to combining tooling, resists and subsequent device processing in a coherent way to achieve a "product" that is economic, has a broad application and is technically productive. Too frequently we develop the technology without regard to the systems aspect (see Figure 1) and consequently do not attain our ultimate goal. We ultimately would like to maximize overlap regions depicted in the Venn diagram of Figure 1. Challenges abound in lithography - extension of optics below 0.5μm, lower cost and higher throughput electron beam systems, economical application of x-rays in the printing process, practical ion beam lithography systems and resists for these methods that have high sensitivity (needed for high throughput), good resolution, and process compatibility.

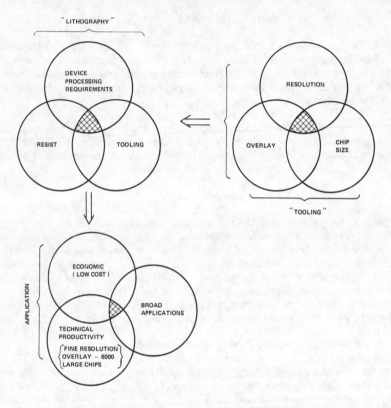

Figure 1 - Systems aspect of lithography.

COMPLEXITY

The complexity (Table I) of semiconductor integrated circuits has progressed from a few gates SSI (small scale integration) to MSI (medium scale integration) to LSI (large scale integration) but not thru VLSI (very large scale integration). The next complexity era, GSI (Giant Scale Integration: .5μm < minimum feature size < 1μm) will be followed by HSI (Horrendous Scale Integration: minimum feature size < .5μm). Many talk of VLSI but few actually practice VLSI in manufacturing - we deal today primarily with LSI and with a few selective parts that truly fall into the VLSI regime. If we examine the trends in minimum feature size, number of active devices, chip or die size and the wafer size, we find that today the wafer area has grown a factor of 25 from the onset of intergration with a soon anticipated growth factor to 40 when 200 mm wafers are in full production. The chip or die size has gone in 20 years from a few square mils to designs at 1 cm^2 with a constantly decreasing minimum feature size and growing number of devices per chip or die. Should these trends continue to sub 0.5 micrometer and beyond, and there is no genuine reason for them not to except for the difficulty of doing manufacturing lithography with large chip sizes and high yields, we will or may find ourselves with an extremely difficult lithography task by the end of the present decade or very early in the 90s. It is not difficult to project a 16 to 64 Mbit dram fabricated in true half micron ground rules having a memory cell of ≤ 10 μm^2 and thus covering an area between 160 and 320 mm^2 or 1.2 cm x 1.2 cm to 1.7 cm x 1.7 cm, where the minimum lithographic image printed in resist is perhaps 0.7 to 0.8 of the minimum designed feature or 0.35 - 0.4 μm.

TABLE I: COMPLEXITY OF SEMICONDUCTOR INTEGRATED CIRCUITS

YEAR	CLASS	NO. DEVICES		LITHO (μm)	CHIP SIZE mm^2	WAFER SIZE mm
60-68	SSI	$2^1 - 2^7$	(2 - 128)	> 10	1-15	25
65-75	MSI	$2^6 - 2^{12}$	(64 - 4K)	3 - 10	10 - 25	50
72-83	LSI	$2^{11} - 2^{17}$	(2K - 128K)	1.5 - 4	15 - 50	50 - 100
80-88	VLSI	$2^{16} - 2^{22}$	(64K - 4M)	.75 - 2	25 - 75	100 - 125
85-93	GSI	$2^{21} - 2^{27}$	(2M - 128M)	.5 - 1	50 - 200	125 - 200
90-99	HSI	$2^{26} - 2^{32}$	(64M - 4000M)	$\leq .5$	100 - 400	≥ 200

Today, no captive or commercial lithographic tool exists that is suitable to satisfy this requirement. Moreover, there are only a few (electron-beam) tools capable of this technical requirement (minimum feature size, overlay, field size) but they lack sufficient throughput for serious manufacturing.

Let us now delve into the individual lithography techniques available to us and explore each one's limitations in hopes of guiding us to a sane conclusion on the method or methods that will satisfy our 16-64 Mbit, 0.5 μm requirement by the end of the decade. There is no doubt that we will be doing 0.5 μm lithography and it is likely that it will not stop there but continue to the sub 0.5 μm regime with 0.2μm needed by perhaps year 2000 (Figure 2). Therefore, in discussions to follow, we must focus not only on the 0.5 μm requirement but also on extension to the 0.2 μm domain. Figure 2 illustrates minimum dimension vs year of production with some extrapolations on how those minimum features may be generated in the 1984 - 2000 era: refractive or reflective optics, E-beam and/or storage ring or other bright source x-ray lithography.

Key issues in lithography are:

- Resolution
    - fine line/pitches with good feature size control
    - alignment capability commensurate with the resolution
- Field or chip size
    - growing larger and larger

Throughput
- a manufacturing requirement

The last figure of this paper shows the ultimate type of single layer resist image we would like to have: good submicron resolution combined with high aspect ratio resist profile and good linewidth control.

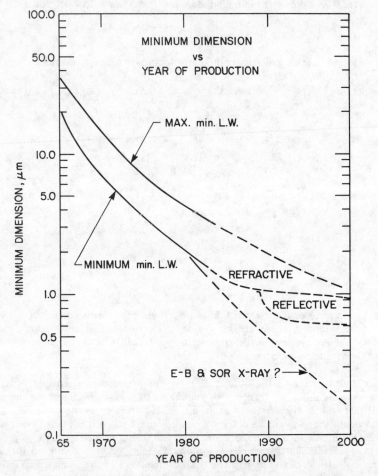

Figure 2 - Image minimum dimension vs. year of production.

LITHOGRAPHY CHOICES

Optical lithography enjoys an advantage of being the major technology in use today in manufacturing. It has high throughput, relatively low cost and the disadvantages of limited depth of focus and resolution and alignment over large chips. Electron beam lithography enjoys the advantages of demonstrated submicron resolution and alignment but disadvantages of low throughput and high cost. X-ray lithography has the potential of high resolution and throughput with some concerns about alignment and cost.

The lithographic techniques available to us for generation of the IC pattern features are:

- Photon exposure (optical wavelengths, $\geq$ 200 nm)

    - 1:1 scanning projection; full wafer printing
    - Nx reduction; chip by chip printing
    - Unique optical systems

- Electron beam exposure

    - Direct write; Gaussian vs. shaped beam
    - Proximity writing (requires masks)
    - Reduction optics (requires masks)

- X-Ray exposure

    - 1x proximity printing; full wafer vs. significant fraction of a wafer printing
    - Weak source vs. bright source

- Ion beam exposure

    - Direct write; focused ion beam
    - Proximity printing (requires masks)
    - Reduction optics (requires masks)

We will next explore some of the expectactions of the limit or critical problems in each of these lithography choices.

OPTICAL LITHOGRAPHY

RESOLUTION

Optical lithography[1] has been the mainstream process for high volume IC manufacturing. However, as we approach the one and submicron feature sizes, we must look critically at the problems encountered in printing these features. The useable optical resolution (R) as a function of numerical aperture (NA) of a lens and as a function of the wavelength (λ) of the illumination is

$$R \simeq \frac{3\lambda}{4NA} \tag{1}$$

The "depth of best focus" (D) for a lens can also be expressed as a function of NA and λ and is

$$D = \frac{\lambda}{2(NA)^2} \tag{2}$$

and eliminating NA gives depth of focus in terms of λ and resolution:

$$D \simeq \frac{R^2}{\lambda} \tag{3}$$

From equation (3) we see that the optical depth of focus decreases as the **square of the resolution and increases linearly with a reduction in wavelength**: Figure 3 illustrates this and we see that at a minimum feature of 1.3μm and 436 nm illumination (reasonable operating parameters for present day optical steppers) the depth of focus is about 4μm. However, as we go to 0.75μm resolution and 365 nm illumination the DOF decreases to about 1.5μm. At 0.5μm, using 250 nm illumination, the DOF is only 1μm.

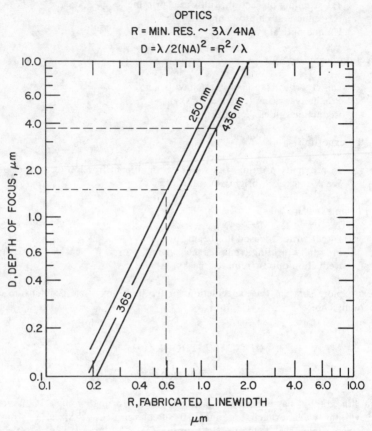

Figure 3 - Depth of focus for optical systems.

D.G. Goodman[2] has recently computed irradiance contours for simple images formed by microcircuit lenses as a function of focus, aberations and linewidth. Figure 4a shows the irradiance distribution for perfect focus of a lens printing a linewidth of 0.75λ/NA. If the required linewidth is decreased to .5 λ/NA, Figure 4b shows a markely poorer irradiance distribution. This illustrates why most people consider the practical lens resolution limit to be .75 λ/NA. Figures 4c and 4d show the disastrous effects of a .5 λ defocus and astigmatism error on the irradiance distribution, respectively. I think it should be clear from Figure 4 that the reliable printing at the lens resolution limit is indeed a dificult task. Doing this over large fields is more difficult.

While no clear formula can be given it is generally observed that the number of printable pixels in a single optical field is on the order of 10^8. Figure 5 illustrates this observation for not only optical but shows the corresponding situations for e-beam and x-ray systems as well. This implies that the maximum field size (in units of resolution element dimension squared) (for a single optical field with no mechanical motion linking or stitching fields together) is about 10^8 x resolution^2. At 0.5μm resolution, the maximum

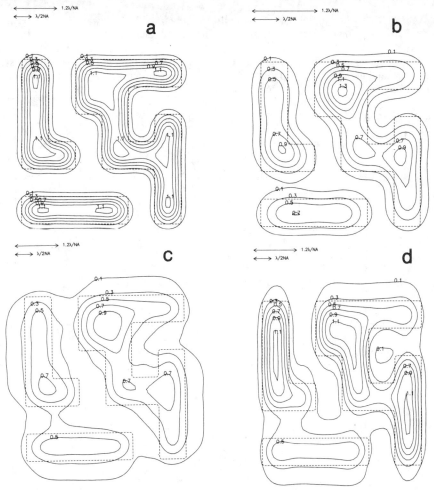

Figure 4 - Irradiance contours for an optical system: (a) linewidth = 0.75 λ/NA; (b) linewidth = 0.5 λ/NA; (c) defocus = 0.5λ; (d) astigmatism = 0.5λ

single field chip size is thus about 25 mm^2 and from Table I we see that this is likely to be too small to print the size of chip needed at this resolution in the time frame required. To print such a chip requires an optical system, including alignment, that can join fields to mechanically multiply the resolution of the basic optical lens. This concept is what Perkin-Elmer accomplished in their Microlign^R series of 1:1 scanning projection printers. These systems, as presently configured, are limited to a resolution of ~ 1.1μm.

There are, therefore, several key concerns about the extendability of optical lithography to the submicron region. They are:

- Quadratic reduction in depth of focus with resolution - nearing the practical mechanical limit (~ ± 1/2 μm.)

- Necessity for mechanical stitching of optical fields to create a single field of adequate size for future chips.

- A ~ 200 nm lower limit for practical optical wavelengths because of inadequate transmission of optical materials below this wavelength.

To get down to the 0.5μm minimum feature size and smaller will require optical systems that are unique and not merely extensions of existing designs if they are to print not only the fine lines but also over fields large enough to cover full chips. Such systems do not exist today at the 0.5μm and below resolution.

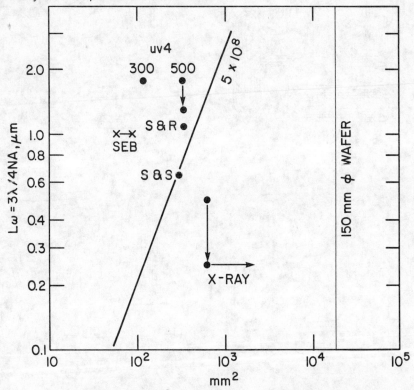

```
SEB = shaped electron beam
S&R = step and repeat optics
uv4 300-500 microlign sys - arrow down is going to uv2
S&S = concepts only - deep uv high N.A. steppers
XRA = storage ring with 4 cm² or larger mask
```

Figure 5 - Basic system resolution and field size without mechanical motion.

Standard multilayer resist systems do not actually improve in anyway the fundamental resolution of the optical systems, but rather make the substrate appear to be more planer. The recent development of a resist that bleaches in the exposed area and thus acts a "contrast enhancement filter" is a development which is likely to assist in the extension of optical lithography to 0.5μm in manufacturing.[3] If thin top resist imaging layer is used, then the MLR effectively separates the imaging and high aspect ratio resist profile functions into two steps. The resist profile is created by RIE or other image transfer methods. These multilayer resists, do however, improve the utility of optical systems greatly in semiconductor IC fabrication near the resolution limit of the tool. It also permits the elimination of deadly standing-wave interference-reflection effects by the addition of absorbing dyes in the planarizing layer.

While optical lithography is the work horse of the industry today, it is facing an uncertain future as we move into the submicron domain for the printing of real chips. If it were possible to decrease the actual chip size as the resolution shrinks then the future of optical methods is more certain but this has not been the trend. Perhaps this trend will change which would then influence the architectural design of computer hardware: chips and packaging. Alternatively, clever optical designs combined with advanced mechanical systems may keep optics growing with the trend line. We will next discuss some radically different lithography techniques: electron beam and x-rays.

ELECTRON BEAM LITHOGRAPHY

SYSTEMS

Electron beam lithography[4] has enjoyed about a decade of use in microcircuit engineering and fabrication. E-beam was, at one time early in its development, predicted to replace optical lithography by now. This prediction failed to materialize because of incorrect evaluations by some which suggested optics was finished at 2 or so microns and overly optimistic throughput expectations for electron beam.

Electron beam lithography has found some important niches which are 1) mask generation: 1x, Nx optical and x-ray, 2) advance circuit development at submicron dimensions and 3) personalizations of masterslice (not necessarily at small dimensions) and 4) selective patterning of small features in a mixed mode operation (i.e., fine gate definition on wafers mostly printed on an optical stepper).

There are at least three problems germain to the question of E-beam lithography taking over the major manufacturing of IC's from optical lithography: 1) system complexity and cost, 2) proximity effects compensation (electron interaction-scattering - with the resist and substrate), and 3) system throughput.

On the first issue, electron beam systems are by their basic nature and the fact that they are pattern generators rather than merely "replicators", significantly more complex and costly in capital and operating supporting dollars as compared with optical lithography tools and this trend seems to be continuing today.

PROXIMITY EFFECT

The second issue, proximity effect compensation[5], has been examined by many and is now quite well understood but not necessarily practical in some cases or practiced at all in others. It has been observed that for 20 KeV electrons and $0.5\mu m$ ground rule devices some form of proximity correction is essential to maintain linewidth control. The model used to explain proximity effects was offered by Chang[5] nearly a decade ago and is the sum of two Gaussian functions: one Gaussian represents the forward scattered electrons or beam spreading and the second Gaussian represents the backscattered electrons from the resist and substrate.

Figure 6 illustrates the range of scattered electrons at 10, 20 and 50 KeV caused by a pencil of electrons injected into 1μm of resist on a silicon substrate.[6] From these Monte Carlo computations it is clear that circuit shapes adjacent to each other are going to each receive some scattered dose (intershape proximity effect) which is a function of size and closeness of all shapes surrounding the shape in question. In addition to intershape effects, there is an intrashape proximity dose which is the dose scattered into the shape while it is being exposed.

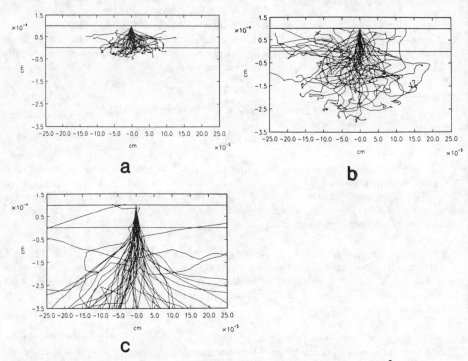

Figure 6 - Monte Carlo simulations of scattering of electrons for 80 $\mu C/cm^2$ dose in 1μm of resist on thick silicon substrate at: (a) 10 KeV; (b) 20 KeV; (c) 50 KeV electron energy.

Several techniques have been explored and utilized to reduce the effects of the scattered electrons. The first, and perhaps the most obvious would be to calculate the total dose received by each shape - intra and inter scattered and direct - and then to adjust the actual exposure dose for each shape so that the resist is given unity, or close to unity, exposure everywhere. This is presently done on IBM Vector Scan and EL-3 electron beam tools for submicron devices. What is not possible, however, is the "removal" of electrons scattered into the regions not being intentionally exposed and hence the spaces between adjacent shapes receive some unwanted dose that cannot be compensated for by the exposure process. Figure 7 shows the absorbed dose as a function of distance into the resist and for 10, 20 and 50 KeV electrons. Here we can easily see the loss of contrast at low electron energy and the decrease in sensitivity at the resist at high energy. Another approach to proximity effect compensation is the use of thick inert resist films to "float" the "imaging layer" above the substrate and this decreases the the number of electrons scattered by the substrate into the imaging regions.

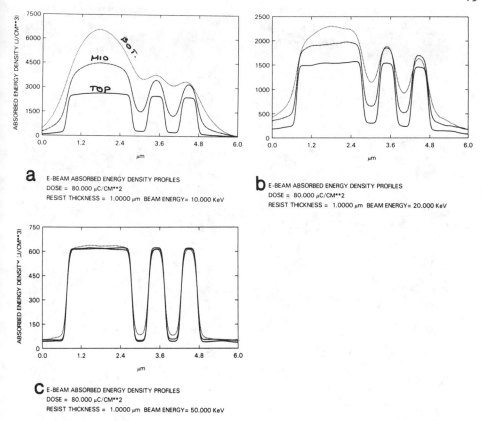

Figure 7 - Cross sectional view of absorbed electron dose as a function of electron energy: (a) 10 keV; (b) 20 KeV; (c) 50 KeV, and as a function of position in the resist layer: top, midway and bottom (Si-resist interface) for pattern of parallel lines, 0.5µm minimum width and larger.

Computer times to determine the best dose for each shape can be excessive as the chip grows in size. Considerable effort has been placed on this problem and the work by Gerber[7] and Jones[8] are notable examples.

Figure 8 illustrates a 0.5µm feature exposed in resist at 25 kV without and with proximity correction implemented with dose control. It is quite apparent that it would be very difficult to delinate, with any reasonable degree of linewidth control, groups of these small features containing isolated shapes, some in close proximity to each other and some adjacent to large shapes without adequate compensation from electron scattering effects.

Use of multilayer resist systems may make proximity compensation easier particularly when the energy of the electron is also selected for optimum conditions. Energy of the electron at 25 keV may be in a region which is not optimum for proximity effect correction. At 10 keV the electrons do not have sufficient energy to penetrate through thick layers of the planerized resist (inert resist layer) to the silicon substrate and consequently backscattering of electrons is not very great. Thus, if a top image layer is used which rests on a micrometer or so of inert resist, proximity effects can largely be ignored. However, this combination has not been very useful in direct patterning of IC's because the primary beam cannot penetrate the resist layers and be scattered off the alignment marks.

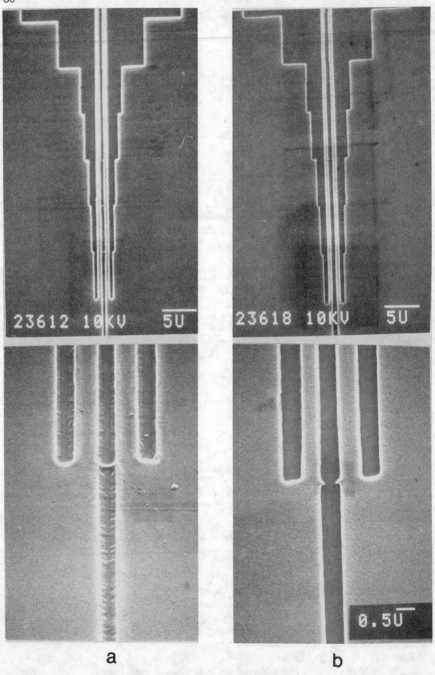

Figure 8 - Example of 0.5μm feature (gap, line, gap) exposed in a "tower" pattern (different size shapes on either side) without (a) and with (b) proximity correction at 25 KeV.

If the resist is removed from the marks by pre-exposure or abalation, then 10 keV can be attractive for device making. In mask making, it is a good combination because no alignment is necessary.

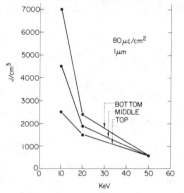

Figure 9 - Absorber dose through the resist as a function of beam energy.

Figure 9 illustrates the absorbed dose through the resist layer as a function of the primary beam energy. Below about 40 KeV the absorbed dose varies significantly through the resist layer and this phenomonenon effectively lowers the contrast at the lower primary beam energies. Computing a figure of merit of the resist/e-beam exposure process further illustrates the advantage of high energy beams. Figure 10 shows the ratio of resist sensitivity and contrast. At lower energies the resist is very sensitive (high absorbed dose) but poor contrast. At high beam energies the resist is two to three times less sensitive but the "contrast" is high. For fine line definition the high energy probe is attractive.

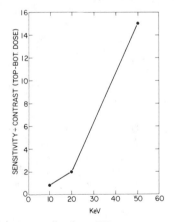

Figure 10 - Resist sensitivity normalized to difference in top minus bottom layer sensitivity for 0.5μm line: i.e. sensitivity ÷ a contrast function.

Throughput: Gaussian vs. Shaped Beam Systems

One serious concern about electron beam systems for manufacturing has been the throughput of these systems. For round gaussian type beams the throughput is inversely proportional to the product of square of the beam stepping distance and dwell time (resist sensitivity). For fine-line fabrication it is common practice to draw a minimum feature line by several passes of a beam: typical 4 to 8 and thus definition of a minimum feature may require from 16 to 64 exposure pixels. For round gaussian type systems using a bright electron source such as LaB_6 or field emission, the system throughput is usually limited by the digital/analog deflection system and not entirely by the resist sensitivity. For shaped beam systems the opposite is usually the case because the beam current density is generally lower.

At what minimum feature size does the throughput of a variable shaped electron beam system merge or equal that at round Gaussian systems? If the average shape written is composed of 4 to 6 minimum feature sizes, for example a $0.5\mu m$ by $2.5\mu m$ line (0.5 x 0.5 μm^2 minimum feature size) then the average number of gaussian spots exposed in one "flash" of the shaped beam system is between 80 and 320 depending on how many beam passes are used to expose the minimum feature shape. Assume five pass and the pixel count is then 125.

The trend in electron beam system design has been to use high current density probes for round gaussian beam systems operating a high stepping rate and to use lower current density shaped spots stepping at a lower rate for variable shaped beam systems. Let the variable shaped beam system have a current density J_{VSB} and the round gaussian beam system a current density J_{RGB} and further assume that the average shape can be exposed as $N_{mfs/shape}$ minimum feature sizes (mfs) for either system where the round gaussian beam system (RGB) would compose a minimum feature size by exposing $N_{spots/mfs}$ discrete spots at a frequency f_{RGB} and the variable shaped beam system (VSB) would expose this average shape by a single exposure at a frequency f_{VSB}. For a resist of sensitivity, S, the shape exposure time for both systems, ignoring all overheads is:

$$T_{RGB} = \frac{N_{spots/mfs} \times N_{mfs/shape} \times S}{J_{RGB}}$$

and

$$T_{VSB} = \frac{S}{J_{VSB}}$$

equating the exposure times and letting $N_{spot/mfs} \times N_{mfs/shape} = n$ we get

$$J_{RGB} = n J_{VSB}$$

where $n = N_{spots/mfs} \times N_{mfs/shape}$. That is, the current density of the round Gaussian beam system will have to equal n times that of the shaped beam system: n being over 100 can place round Gaussian beam systems at a disadvantage from a throughput point of view - but that is not the whole story. One other significant consideration is that shaped beam systems have been built and are operating in the ≥ 5 Mhz step rate range. The round Gaussian beam system will have to operate at n times the step rate of the shaped beam system.

Note that the actual stepping rate is determined by the ratio of resist sensitivity and beam current density. To make a system step faster requires either a faster resist (which may have poorer resolution) or a higher beam current density (which tends to have a large spot or probe size).

For a Gaussian system to compete, assuming the high current density can be achieved without enlargement of the spot, it is possible that they have to step at a rate well in excess of several hundred megahertz - this is difficult to achieve. If the RGB system has a current density near the limit of about 300 A/cm^2 for a $0.1\mu m$ spot then with a fairly agressive positive resist having a sensitivity of $5\mu C/cm^2$ the maximum stepping rate is

$$f = \frac{300 A/cm^2}{5\mu c/cm^2} = 60 Mhz$$

To operate the system any faster would require a resist significantly faster and below a sensitivity of about $0.5\mu C/cm^2$ we begin to get concerned with the number of electrons per pixel or spot.

Throughput Reduction with Minimum Feature Size

We have noted in experiments of minimum feature size reduction with high pattern quality (reduction factor r) on a shaped beam system (modified from 1 μm minimum features) that it was necessary to decrease the beam current density by a factor almost equal to the reduction factor, and also to decrease the maximum size shape by the same reduction factor r. To maintain the same image fidelity, as the feature size reduced, it is also necessary to decrease the beam edge rise distance to maintain a constant ratio of minimum feature size to beam edge rise distance. The implications of these observations are: the time to write a given chip with a shaped beam system increase as the chip minimum feature sizes decreases because the beam current density must be decreased to maintain adequate edge size control. In Gaussian systems this may not be as necessary.

What is the throughput of electron beam systems as the minimum feature size is decreased from the 1 micrometer domain to sub 0.5 micrometer? Does the throughput in wafers/hour remain constant? Several interesting observations have been made.

When the minimum feature size is reduced a factor r, the number of pixels to write per chip or die stays constant, if no proximity correction is performed. The **chip thoughput** will thus stay constant but the wafer throughput will decrease by $\sim r^2$. For minimum features at and below a micrometer, the shape or exposure pixel count will grow (from that of minimum feature $\geq 1.5\mu m$) a small amount because some shapes will have to be partitioned to optimize the proximity correction. This may or may not be a small reduction in throughput.

In shaped beam systems, the electron-electron interaction causes a spread in the energy of the incoming electron beam as well as its trajectories which adversely affects the quality of the shaped beam image. Second, small column charging and scattered electron effects degrade the image and have to be reduced. To reduce these effects it is necessary to simply reduce the number of electrons in the beam and hence the current must be decreased. The total current is automatically decreased for a constant current density because we note that as the minimum feature is reduced, so was the **maximum** feature size, i.e. if at $1\mu m$ mfs it was possible to draw a 4 x 4 $(\mu m)^2$ shape then at $0.5\mu m$ mfs it is likely that only a 2 x 2 $(\mu m)^2$ can be drawn (maintaining the same image quality). But what we also point out is that it was necessary to also **decrease the current density** by a factor between r and r^2 (perhaps $r^{1.5}$). This is a significant unexpected chip and wafer throughput reduction of a shaped beam system with shrinking feature sizes. When the maximum shape size and then the current density was decreased the shaped aperture image quality became adequate for small feature size generation.

Thus the wafer throughput can be expected to decrease at least by r^{3-4} as the minimum feature size is decreased by a factor r (r = 1/2 for mfs going from 1 to .5μm). One can try to write faster as the feature size is decreased to compensate for the throughput reduction effects, but if the same resist is used, it will be necessary to increase the beam current, but this can only be done at the expense of image qaulity and usually we find it necessary to go the other way and decrease the step or clock rate to allow more settling time for the electronics and deflection systems. This can cause another throughput reduction by a factor of r.

Lastly, resist tends to require more dose as we shrink the minimum feature sizes for the same quality image. Perhaps this is another factor of r^{1-2}. If we are near the minimum of ~ 200 electrons per resolution pixel then the throughput is reduced by r^2. This is necessary to give a minimum number of electrons to a pixel element. To summarize at a constant wafer throughput:

EFFECT	IMPACT ON THOUGHPUT
Pixel count growth	r^2
Current density effect (shaped beam system)	r^{1-2}
Less sensitive resist for improved resolution	$\sim r^{1-2}$
Deflection clock settled/noise reduction	$\sim r^{0-1}$
Wafer growth	r^2

Hence the wafer throughput of a shaped electron beam system can be expected to decrease a factor of between r^4 and r^9 as the minimum features decrease in size. Scanning shaped electron beam systems designed for a current density of 50 amp/cm^2, .2 µm beam edge rise, 1.0µm minimum feature, 4 x 4 µm^2 maximum shape and 30 wafers/hour for a specific VLSI chip may be able to only write 30 ÷ 16, or perhaps as little as 30 ÷ 32 (or less), one wafer/hour, at 0.5µm ground rules. Note that the beam current varies a factor of 32 times as the size of the shape that is written is changed (from 16µm^2 to .25 µm^2). It is not entirely clear to us at this time if these are throughput degrades can be removed by new design concepts or implementations.

Some suggestions are to maintain as large a maximum shape size as possible as the minimum feature shrinks. This can be achieved perhaps by reducing electron-electron interactions (i.e. minimum tight bunching or confinement of the electrons as they traverse column) via new column designs. Such a design may also allow increased current density instead of having to reduce it for submicron systems. If these objectives can be attained, then we might find experimentally that a variable submicron shaped beam system can be built that has a reasonable throughput.

X-RAY LITHOGRAPHY

A full discussion of x-ray lithography is well beyond the scope of this talk and article. We are going to explore x-ray lithography from the point of view of bright sources such as a storage ring.[9] I will assume that the engineering probelms of masks, alignment etc. can, with adequate attention, be solved. The question is - what is the expected lower limit in resolution for x-ray lithography?

RESOLUTION

Proximity printing with x-rays is not a new technology. What is new is the potential use of this method for IC production using bright x-ray sources. In this case it is expected that the resolution of the process will be limited by diffraction and, perhaps, secondary electrons. Figure 11 illustrates the printing process with collimated x-rays from, for example, a storage ring incident on a mask with a 0.5μm slot. The mask/wafer gap is shown as 40μm. For those who have done optical contact printing, this situation, in terms of diffraction, is equivalent to printing 0.5μm lines with "vacuum contact" of 0.1μm at optical wavelengths.

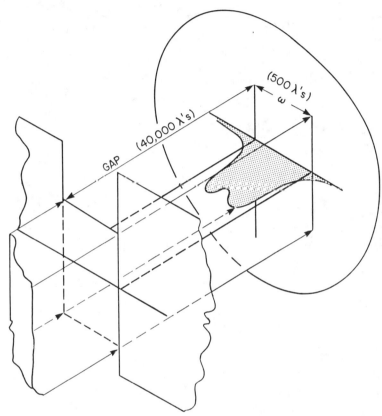

Figure 11 - X-ray proximity printing process. Distance given in units of x-ray wavelength: 0.5μm line at 40μm mask/wafer gap. 10Å

What we want to know is how fine a line can we print as a function of mask/wafer gap. Figure 12 illustrates the diffraction of x-rays from a storage ring operating at 750 meV, 1.9 meter bending radius magnets, transmitted by a Beryllium window, and silicon mask. The absorber is considered perfect.

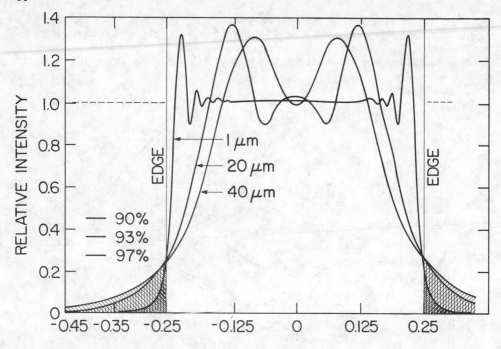

Figure 12 - X-ray diffraction for Brookhaven VUV storage ring x-rays passing through a silicon mask as a function of mask/wafer gap.

To answer how fine a line can be printed. An effective "modulation transfer function" is computed for x-ray lithography by determining after diffraction the percent dose remaining within the fine line width. Figure 13 illustrates the percent dose incident on the wafer in a given width line as a function of mask/wafer gap. (These curves are determined by calculating fresnel diffraction for the continum of storage ring wavelengths at 3 specific gaps and then integrating the flux incident within the specified linewidth).

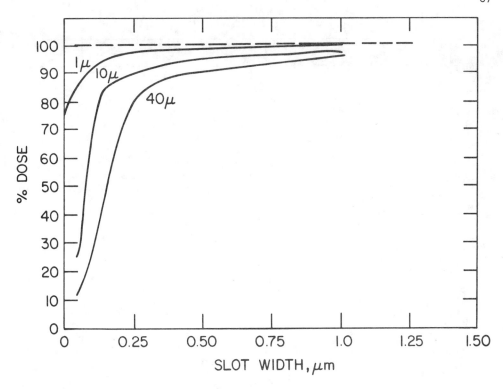

Figure 13 - Flux in a line width defined by the mask integrated over the storage ring spectrum transmitted by the silicon mask. This might be considered an x-ray "MTF" curve.

At 40μm gap and a 60% dose level, at least 0.2μm line can be printed. If the gap is reduced to 10μm, about the lower practical limit for mask/wafer gap, then it is expected that sub 0.1μm lines could be printed. These computations on lower limits for x-ray lithography do not take into account secondary electrons, however, the range of these do not exceed about 0.02μm, well below the limits imposed by diffraction at practical mask/wafer gaps.

We conclude that if x-ray lithography can be introduced for IC production at perhaps 0.6 - 0.5μm then it can be expected to enjoy a usefullness until sub 0.1 μm minimum feature sizes are reached in production.

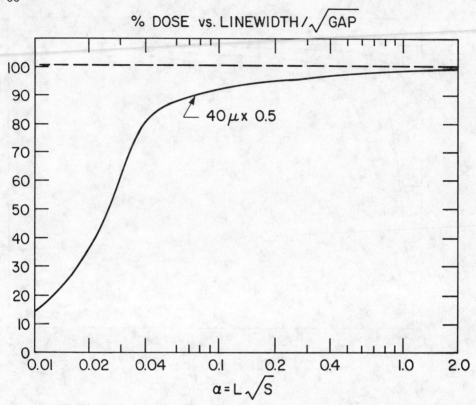

Figure 14 - Generalized version of Figure 13, percent dose vs. mask/wafer gap.

Figure 14 is a generalized version of Figure 13, percent dose vs. linewidth for any gap. Figure 15 is a typical example of an x-ray exposed resist profile with vertical sidewalls in thick resist combined with submicron resolution.

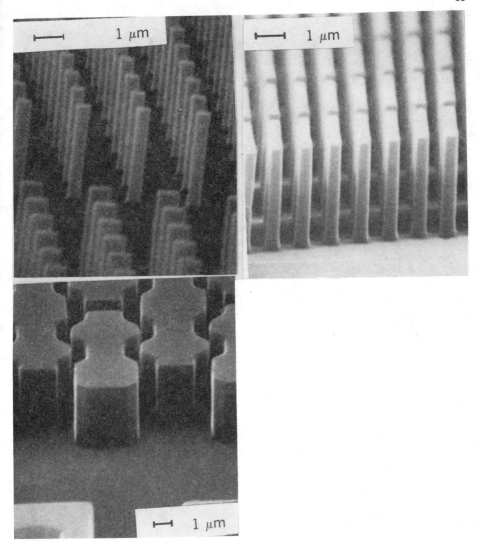

Figure 15 - X-ray exposed 3.4µ thick PMMA resist from IBM Synchrotron radiation beamline at the National Synchrotron Light Source, Brookhaven, NY.

SUMMARY

In summary, I would like to leave you with the following thoughts about optics, e-beam. x-ray and ion beam lithography for the next 5 to 10 years.

REFERENCES

1. B.J. Lin, Optical Lithography in Fine Line Lithography, Ed. R. Newman, North-Holland Co, 1980.

2. D. Goodman, private communication, to be published.

3. B. Griffing and P. West, Spie Conf. No. 394, p. 33, Ed. H. Stover, March 1983.

4. A.N. Broers and T.H.P. Chang, Microcircuit, Cambridge University Press, Ed. Ahmed and Nixon, 1980.

5. T.H.P. Chang, J. Vac. Sci. Tech., Vol. 12, p. 1271 (1975).

6. M. Rosenfield, private communication, to be published.

7. W. W. Molzen and W. D. Grobman, J. Vac. Sci. Technol., 19(4), Nov./Dec. 1981; and P. D. Gerber, W. W. Molzen, to be published. Presented at 10th International Conference on Electron and Ion Beam Science and Technology, May 9-14, 1982, Montreal, Canada.

8. M. Jones, (this conference)

9. W.D. Grobman, Handbook on Synchrotron Radiation, Vol. 1B, Ed. E. Koch, North-Holland Co., 1983.

OPTICAL

- Good throughput and low cost
- Limited resolution and depth of focus
- {Limited chip size ($\sim 10^8$ pixels) }$^+$

E-BEAM

- Resolution good (high KeV or MLR)
- Very questionable throughput for mfs $\lesssim 0.5 \mu m$ and costly.
- {Limited field size, table stitching}$^+$

X-RAY

- Good resolution
- Good throughput
- New technology - risks/costs

IONS

- Not for lithography (general)
- Good for micromachining

+ Present engineering problems

ACKNOWLEDGEMENT

I would like to acknowledge the contributions of D. Goodman and M. Rosenfield - optical irradiance distributions and e-beam proximity effect simulations, respectively. Helpful discussions with R. Acosta, A. Broers, P. Coane, B. Lin, J. Maldonado W. Molzen, J. Warlaumont, R. Viswanathan on lithography issues are acknowledged with technical assistance of H. Voelker and L. Towart and P. Bailey.

CORRECTION OF ELECTRON BEAM PATTERNS FOR EXPOSURE AND PROCESS EFFECTS

P. Hendy, M.E. Jones, P.G. Flavin, and C. Dix
British Telecom Research Laboratories, Martlesham Heath, IP5 7RE, UK

ABSTRACT

A program for correction of proximity effects which occur in electron beam lithography has been developed for a DEC PDP 11/44 computer for use with a Cambridge Instruments EBMF system. Sequential corrections are made for both intra- and inter-proximity effects, using data obtained from look-up tables. The tables are generated from the results of test pattern exposures and can be readily changed for different resist/substrate combinations. Allowance may also be made for the effects of subsequent processing steps. Application to device development programmes have included intra-corrections to generate resist profiles with controlled undercut for metal lift-off, and linewidth bias for the wet etching of oxide windows. Inter-corrections have been used for lift-off of 1um thick Ti/Au patterns.

INTRODUCTION

The use of electron beam lithography in VLSI circuit development offers advantages in resolution, overlay accuracy and flexibility. However, electron backscattering from the substrate results in a non-uniform exposure distribution within pattern features and leads to undesirable variations in the final resist image[1]. These variations, depending on both the size of individual shapes and the mutual scattering between shapes, are termed intra- and inter-proximity effects respectively. They are particularly important when pattern dimensions are comparable with the electron scattering range, as occurs in VLSI device structures.

Corrections to compensate for proximity may be effected by a simultaneous solution of the dose values[2-5], so that all shapes receive an equal average exposure intensity. However, such techniques require a knowledge of the exposure intensity distribution (EID)[1], and substantial computing resources to solve the large sets of self consistent scattering equations. Several other methods which have been developed[6-9] also require large computing resources.

In contrast to these methods the technique described here derives pattern corrections from look-up tables, which are easily constructed from the results of test exposures[10], and runs on a DEC PDP11/44 minicomputer. Pattern processing is sequential, with intra-correction applied by dose and optional shape size modification; and inter- correction applied by shape fracturing and exposure dose adjustment.

CORRECTION PROCEDURES

The correction program was written for a Cambridge Instruments EBMF to operate on the primitive shapes of the exposable data, although the techniques are general for any vector scan system. During the intra- phase each primitive shape is selected sequentially, its true size identified by establishing whether it forms part of a larger shape, and this size used to determine the required dose adjustment and shape bias from the look-up table[10]. Any compensation for process induced dimension changes may also be made at this stage[11].

0094-243X/84/1220092-06 $3.00 Copyright 1984 American Institute of Physics

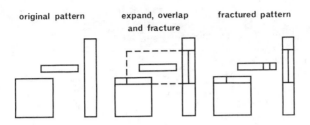

Fig.1. Expand, overlap and fracture of shapes for inter-correction.

After intra-correction the pattern data has assigned dose values, so that during the following inter-correction phase compensation is made specifically for the inter-shape scattering. Correction needs to be made over what may loosely be termed the effective scattering range (ESR), the range over which most of the electron energy is deposited and the equivalent of β in the EID function. The size of each shape is increased by an amount equal to the ESR and any shapes overlapped by this expansion are fractured. This simple operation is illustrated in Fig.1. Non-overlapped portions retain their original dose, whereas shapes formed by the overlap are given a modified dose calculated from the product of a value from the look-up table with two scaling terms. The look-up table value is determined by the separation of the shape pair, with the scaling terms proportional to the sizes of the overlapping shapes and their unmodified doses.

PROGRAM OPERATION

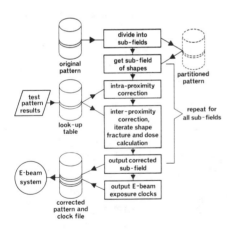

Fig.2. Flow diagram of program.

The overall structure of the program is shown in Fig.2 together with the inputs and outputs. The data is divided into 10 bit subfields for correction to reduce the number of tests required for mutual scattering between shape pairs[2]. Shapes within the ESR of a common subfield boundary are identified and included in the adjacent subfield for processing. Each subfield is corrected in turn, first for intra-proximity, followed by inter-proximity, with any optional iterations. Fractured shapes are regrouped where appropriate, and are assigned to one of the 16 available clock numbers. These are set by the output clock file at exposure time to determine the dose values.

The program source language is FORTRAN and requires 31k words of memory with the input and output sections overlaid. Test patterns containing random arrays of rectangles distributed in size and separation between 0.5µm and 20µm have been used to assess the performance of the program. Results from these, shown in Table I, were obtained on a PDP11/44 system with RL02 disks. The ratio of output to input shapes increases with the reduction of lithographic dimensions, and is 3.3 for patterns in which 50% of the shapes have 0.5μm separation. Processing time is comparable with other system utility programs for pattern processing.

Table I Correction program performance statistics

PATTERN NUMBER	1	2	3	4
Percentage of shapes at 0.5μm separation	10	50	20	20
Shape count before correction	1409	1536	3538	10078
Shape count after correction	2442	5024	8799	23990
Shape count ratio after/before	1.7	3.3	2.1	2.4
Correction time (sec/1000 original shapes)	55	71	91	72

LOOK-UP TABLE OPERATION

Look-up tables, experimentally derived from calibration patterns are used for both intra- and inter-correction. Tabular information for intra-correction is derived from test exposures of lines of varying widths, each width being exposed at a range of dose values, and linewidth measurements being made to establish the optimum dose for that width. Example results, normalised to a large area exposure dose, are shown in Fig.3 for PMMA on Si and GaAs substrates.

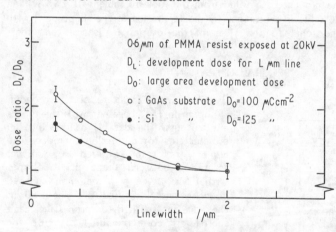

Fig.3. Calibration pattern results for intra-proximity look-up tables.

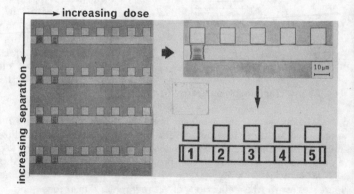

Fig.4. Calibration pattern to measure the inter-shape scattering dose.

The enhanced intra-proximity effect on GaAs which results from its higher atomic number can be seen. The look-up table can also be derived from cross sectional data of resists when close control of the profile is required eg. for lift-off. However this can lead to a loss of linewidth control unless bias is applied simultaneously[11].

Calibration patterns for inter-proximity correction consist of a matrix of pairs of squares, having constant separation along each row and constant dose in each column as shown in Fig.4. The top line of squares in each row is exposed at the reference dose, whereas the bottom line alternates between the reference dose and a variable dose that is incremented along the line. The square opposite each reference dose receives a contribution due to inter-shape scattering. When the resist boundary of the lower line continues unperturbed between two reference squares the inter-shape scattering dose is determined as the difference between the reference and variable doses at that location. The amount of inter-scattering dose as a function of shape separation is plotted in Fig.5, again showing the increased scattering from GaAs.

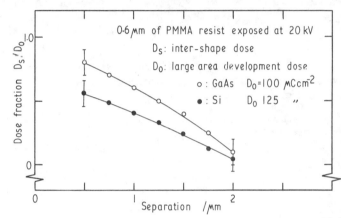

Fig.5. Calibration pattern results for inter-proximity look-up tables.

APPLICATIONS

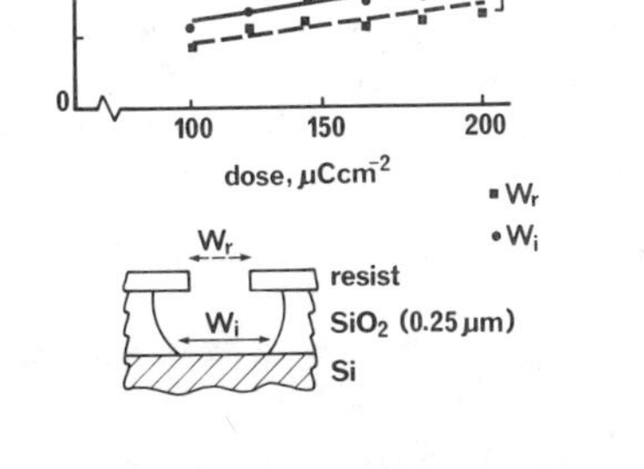

This flexible correction technique has been used in several device development programmes. Intra-correction has been used to obtain reproducible resist undercut profiles in PMMA for metal lift-off on GaAs[11] and Si substrates at linewidths down to 0.25μm. At small device dimensions in Si the use of wet processing for oxide etching is usually discounted because the isotropic process inevitably introduces unwanted lateral etching. However, by compensating for this dimensional variation as an additional bias in the intra-correction routine, submicron linewidths at the Si_2 interface have been wet etched for ECL ULA's[12]. In this process a

Fig.6. Resist and oxide window width versus exposure dose.

0.2μm line was exposed in resist with wet etching of a 0.25μm thick oxide layer resulting in a 0.6μm window. The relationship between resist linewidth and oxide window width as a function of dose are shown in Fig.6.

The results of an inter-correction test pattern on GaAs with feature sizes from 0.5μm to 5.0 μm and 0.5μm gaps are shown in Fig.7. Uniform exposure of the uncorrected pattern at dose values necessary to clear the 5.0μm and 0.5μm features are shown in (a) and (b), with the pattern after proximity correction shown in (c).

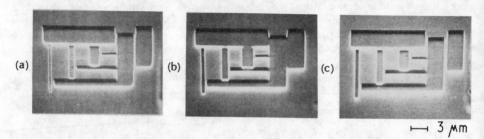

Fig.7. Test exposure of 0.7um PMMA on GaAs at 20kV. (a) uncorrected at 100μCcm^{-2}, (b) uncorrected at 150μCcm^{-2}, (c) proximity corrected.

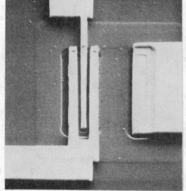

The result of application to the metal interconnect layer of the submicron ECL ULA is shown in Fig.8. The Ti/Au metal is 1μm thick and the separation also 1μm. It was defined by lift-off in a bilevel resist structure but required inter-proximity correction to obtain controlled undercut at this linewidth separation.

Fig.8. Transistor with 1.0μm thick Ti/Au metallisation.

SUMMARY

A proximity correction method has been described that is applicable for both intra- and inter- shape correction, but has only modest requirements for computing resources. Its flexibility of operation results from the use of look-up tables containing the appropriate calibration data, which are experimentally derived from test exposures. The use of these tables has enabled the correction facilities to be extended to include appropriate allowance for subsequent device processing. The processing time and resulting data expansion are comparable with other pattern processing tasks performed on the minicomputer system.

ACKNOWLEDGEMENTS

Acknowledgement is made to the Director of Research, British Telecom for permission to publish this paper.

REFERENCES

1. T. H. P. Chang, J. Vac. Sci. Technol., 12 (6), 1271-1275, (1975)
2. M. Parikh, J. Appl. Phys., 50 (6), 4371-4387, (1979)
3. N. Sugiyama, K. Saitoh, K. Shimizu and T. Tarui, Electronics and Comm., 62-C (10), 88-97, (1979)
4. Y. Machida, K. Nakayama and T. Hisatsugu, Fujitsu Scientific and Tech. J., 16 (3), 99-113, (1980)
5. E. Kratschmer, J. Vac. Sci. Technol., 19 (4), 1264-1268, (1981)
6. N. Sugiama and K. Saitoh, Proc. 9th Int. Conf. on Electron, Ion and Photon Beam Science and Technol., 272-281, (1980)
7. D. P. Kern, Proc. 9th Int. Conf. on Electron, Ion and Photon Beam Science and Technol., 326-339, (1978)
8. N. D. Wittels and C. I. Youngman, Proc. 8th Int. Conf. on Electron, Ion and Beam Science and Technol., 361-370, (1978)
9. A. M. Carroll, J. Vac. Sci. Technol., 19 (4), 1296-1299, (1981)
10. P. Hendy and M. E. Jones, Proc. Microcircuit Eng.'82, Grenoble, 22-27, (1982)
11. C. Dix, P. G. Flavin, P. Hendy and M. E. Jones, J. Vac. Sci. Technol. 19 (4), 911-915, (1981)
12. P. G. Flavin, B. A. Boxall, P. E. Holmes, P. Hendy and G. Ravenscroft Proc. Microcircuit Eng.'83, Cambridge, 439-446, (1983)

Part V.   Reactive Ion Etching

PLASMA-ASSISTED ETCHING: ION-ASSISTED SURFACE CHEMISTRY

J. W. Coburn and H. F. Winters
IBM Research Laboratory, San Jose, California 95193

ABSTRACT

The directional etching capabilities of certain plasma-assisted etching configurations are usually attributed to energetic positive ion bombardment. In some gas-surface systems, (*e.g.*, Si-F), energetic ion bombardment can cause a dramatic increase in the rate of reaction between gas phase species and solid surfaces. However, in other gas-surface systems (*e.g.*, Al-Cl_2), no significant increase in the reaction rate is seen upon initiation of energetic ion bombardment; in fact, sometimes a decrease is observed. In these latter situations, if directional etching is to be obtained, additional condensible species must be present in the gas phase (*e.g.*, CCl_x, BCl_x) which tend to deposit on surfaces forming thin protective layers. The etch rate of these so-called "blocking" layers must be accelerated by energetic ion bombardment. Recent work will be described in which the *primary* role of ion bombardment in ion-assisted surface chemistry is discussed. That is, which of the steps in an etching process (reaction or chemisorption, product formation, product desorption) is *directly* influenced by energetic ion bombardment?

INTRODUCTION

Many etching steps in VLSI circuit fabrication must be anisotropic; that is the pattern transfer must be accomplished without appreciable undercut or etch back beneath the mask layer. Plasma-assisted etching methods can accomplish this anisotropic or directional etching and consequently there has been a rapid growth in this technology (see Refs. 1-3 for recent reviews of this subject). Process development has been largely empirical because of the complexity of the reactive gas glow discharge environment, a lack of in-situ species characterization methods (see Ref. 4 for a recent review of optical diagnostic methods) and little understanding of radical-surface chemistry either with or without energetic particle bombardment (see Ref. 5 for a recent review of this subject). The purpose of this summary paper is to direct the reader to recent research in this latter area and to summarize our views as to implications and questions concerning mechanisms in ion-assisted etching.

Surface chemistry research related to plasma-assisted etching has involved the use of controlled beams of species incident on relatively clean surfaces in ultrahigh vacuum conditions. The controlled beams are selected so as to be somewhat representative of conditions encountered in a reactive gas glow discharge. This beam approach allows careful characterization of surface conditions and primary etch products. Very early work in plasma-assisted etching systems suggested that energetic positive ion bombardment of an etched surface was required for directional etching. Consequently most of the controlled beam studies involve a source of energetic positive ions and an independent source of chemically reactive neutral species. The beam studies lend support to the notion that directional etching requires energetic positive ion bombardment; a conclusion which is accepted by most workers in this field.

In certain chemical systems (*e.g.*, Si-F, W-F) energetic ion bombardment causes a large increase in the rate at which gaseous species react with the solid surface[6-9] and this phenomenon will lead to directional etching in certain situations. However, it is necessary also to consider condensible species (*e.g.*, CF_2, CF, CCl_2,...) which are created in the gas phase and arrive at the etched surface. The removal of these species will also be enhanced by energetic particle bombardment and consequently directional etching can be obtained in gas-solid systems where ion bombardment does not enhance the rate of reaction between arriving gaseous species and the solid being etched (*e.g.*, $Al-Cl_2$).[9-11] A process based on this phenomenon requires careful manipulation of the gas phase chemistry to obtain a condition where the arrival rate of condensible species is adequate to inhibit or block etching on sidewalls (where there is no ion bombardment) but not so large as to significantly interfere with etching of the bottom surfaces which are subjected to ion bombardment. It is often difficult in plasma-assisted etching environments to determine whether ion-assisted etching of the substrate or ion-assisted etching of various arriving gas phase species is the primary reason for the observed directionality, since almost all etching plasmas contain species which can participate in sidewall blocking (*e.g.*, the etching of resists introduces carbon containing species into the gas phase; the presence of oxygen can result in oxidative blocking of sidewall etching, *etc.*).

ION-ASSISTED ETCHING MECHANISMS

Whereas there is reasonably general agreement as to the importance of energetic positive ion bombardment in directional etching, there is no such consensus as to the detailed mechanisms by which ion bombardment influences gas-surface reactions even for the most thoroughly studied chemical system, silicon-fluorine.

Before this subject is discussed in detail, it is appropriate to discuss some phenomena which are not considered by the authors to be important in the context of ion-assisted etching. First, one might inquire as to the role of macroscopic heating of the surface by the ion bombardment. It has been shown, however, that large etch yields are seen with extremely low power density in the ion beam[7,8] and also that the etch yield does not depend on the ion beam current density[12] if the reactive gas flux is large enough to maintain an adequate coverage of the surface with etchant species. Secondly, the role of electronic excitation, which is sometimes mentioned in this context, does not seem to be important. Ions in the energy range below 1 keV incident on a solid generate nuclear motion much more effectively than they generate electronic excitation, whereas electrons generate electronic excitations more effectively than they generate nuclear motion. Therefore, if electronic excitation were an important process for causing enhanced etching, electrons should be more effective than ions. In fact, just the opposite is usually observed.[6] Thirdly, reaction of gas phase species with the "hot spot" created by the incident ion after the ion impact does not seem likely when the spatial and temporal extents of the "hot spot" are considered together with the magnitude of normal reactive gas fluxes. Finally, the importance of a reaction driven by the exothermicity of the gas-surface product formation process can be dismissed in some cases by *sequentially* exposing the surface to the reactive gas and then to the ion beam and observing etch yields[7,12,13] which are comparable to those observed with *simultaneous* exposure to the ion beam and the reactive gas.

It has been emphasized that an etching process can be thought of as three sequential steps:[14]

    A.   Reaction between the gas phase species and the surface, most probably in the form of a chemisorption process.
    B.   Formation of the product molecule.
    C.   Desorption of the product molecule.

Furthermore, it is apparent that if ion bombardment is to cause a steady state increase in the etch rate, the rate of each of these sequential steps must increase. The question that will be addressed here relates to the *primary* effect of the ion bombardment. Much of the discussion will be centered around the silicon fluorine system, which has been used for most of the experimental work as of this time.

The view of the present authors, for silicon surfaces exposed to excess fluorine, is that step B, the product formation step, is the rate-limiting step and the *primary* role of energetic ion bombardment is to accelerate the formation of the final etch product by promoting atomic motion and/or by accelerating the field-assisted penetration[5] of the silicon lattice by the fluorine. X-ray photoelectron spectroscopic studies of fluorinated silicon surfaces[15,16] indicate that the majority of the surface is composed of SiF_x (x=1,2,3). This fluorinated layer is relatively stable, remaining on the surface after the source of gas phase fluorine has been removed. As was mentioned briefly earlier, when a fully fluorinated silicon surface is subjected to energetic ion bombardment *in the absence of gas phase fluorine*, initially very large etch yields are observed[7,12,13] which are approximately equivalent in magnitude to the yields measured in a steady state condition of simultaneous exposure of the silicon surface to the ion flux and the gas phase fluorine. A similar observation has been made by Loudiana et al.[17] on the SiO_2/XeF_2 system. These observations demonstrate that energetic ion bombardment can accelerate steps B and/or C sufficiently to account for the magnitudes of etch yields that are observed in ion-assisted chemical etching. The view of the present authors, that the primary effect of the ion bombardment is to accelerate step B, assumes that the product desorption step C is fast enough to keep up with the accelerated product formation step and is therefore not rate limiting. The combined increase in these two steps B and C will reduce the fluorine coverage on the silicon surface. This, in turn, is believed to cause the required increase in the rate at which gas fluorine reacts with the silicon surface (step A). That is, it seems reasonable that gas phase fluorine will react more efficiently with clean silicon (after the ion bombardment has gasified the fluorinated surface layer) than with the fluorinated surface layer itself which is inevitably present when there is no ion bombardment. This increased "sticking probability" caused by ion-induced cleaning was originally suggested by Mauer et al.[18] in connection with a model based on the assumption that the *primary* role of the ion bombardment is to accelerate the product desorption step C via chemically enhanced physical sputtering of reduced silicon fluorides SiF_x (x=1,2,3). The question as to the relative importance of chemical sputtering (where the formation of the final product is enhanced by the ion bombardment) and chemically enhanced physical sputtering (where the final product exists on the surface with the binding energy lowered by the halogenation[18,19]) is very difficult to answer definitively. The following characteristics of the process should be considered in this regard:

1. The nature of the etch product. The mechanism of chemical sputtering is supported whenever the product species observed in the gas phase are shown *not* to exist on the surface prior to ion bombardment. The data on product species in the ion-assisted etching of Si with XeF_2 ranges from predominantly SiF_4[7,20,21] with some SiF and SiF_2 to predominantly SiF and SiF_2.[22] In the absence of ion bombardment, SiF_4 is the dominant product for both XeF_2 and F atoms.[20,23,24] Since SiF_x (x=1,2,3) are the predominant species in a fluorinated silicon surface,[15,16] the SiF_4 product almost certainly arises from a chemical sputtering mechanism. Moreover, the existence of substantial amounts of SiF and SiF_2 in the product beam is not inconsistent with chemical sputtering in that these species are seen at the 10-20 percent level in the spontaneous (no ion bombardment) etching of silicon with F atoms[24] and XeF_2[20] and therefore can be formed and ejected from the surface by mechanisms other than physical sputtering. Also, it is well known that SiF_2 is the dominant volatile product when the Si surface is heated to high temperatures.[25] Concerns as to how representative Si/XeF_2 reactions are of Si/F atom reactions[26] are not relevant in the context of this discussion because the ion assisted phenomena of interest are those observed primarily in the Si/XeF_2 system. Similar effects have been observed with Si/F_2[12] and it is anticipated that Si/F atom systems will exhibit large ion enhanced effects. Nevertheless, these product distribution measurements are very difficult to interpret quantitatively because of a lack of information on ionization cross sections and fragmentation patterns for the reduced silicon fluorides and uncertainties as to the velocity and angular distributions of the various species. These results are also influenced by the neutral-to-ion flux ratio in the experiments in that excessive ion current density may deplete the surface of adsorbed fluorine thus lowering the possibility of forming SiF_4. Some evidence to support this effect has been presented[12] and a study of the chlorine coverage as influenced by ion bombardment has been made for the $Si-Cl_2$ system.[19,27] Very recently[28] it has been shown in the Si/XeF_2 system that the product distribution can be shifted from predominantly SiF and SiF_2 at low XeF_2/Ar^+ flux ratios to predominantly SiF_4 at higher XeF_2/Ar^+ flux ratios.

2. Energy distribution of products. Recently some important studies of the energy distribution of the product species have been carried out for both

the silicon-fluorine[22] and the silicon-chlorine systems.[29,30] These results are compared with the $E/(E+E_b)^3$ dependence expected from collision cascade theory for sputtering where E is the kinetic energy of the product species and E_b is the binding energy of the species to the surface. This data has a tendency to follow an E^{-2} dependence at high E values, and these results have been interpreted as indicating that a physical sputtering process is dominant.[31] It is our opinion that the E^{-2} dependence is consistent with either physical sputtering, chemical sputtering or detrapping (to be discussed in greater detail later) mechanisms. Any species, regardless of its origin, will have an energy distribution which is influenced by the collision cascade if it is desorbed during the period that the cascade exists. For example, an SiF_4 molecule which was generated and desorbed during a single collision cascade (*i.e.,* chemical sputtering) could have an E^{-2} energy dependence. The SiF_4 could also be generated during one collision cascade and desorbed during the next and the result would still be chemical sputtering with an E^{-2} energy dependence. If exposure of Si to XeF_2 produced SiF_4 which was physically trapped beneath the surface, then ion bombardment could subsequently liberate these molecules and the energy dependence could also be E^{-2}.

Whereas these product energy distribution studies do show an approximate E^{-2} dependence at high energies, the value of E_b obtained from this data is too low (≤ 0.5 eV) for the species to remain on the surface at room temperature. Such a low binding energy is consistent with either chemical sputtering or ion induced detrapping but is not expected if only physical sputtering were involved. Again it should be emphasized that there is evidence that the cross-section for removal of fluorine from a fluorinated silicon surface is dependent on the fluorine coverage,[12] ranging from cross sections similar to those observed in the physical sputtering of chemisorbed gases at low fluorine coverage, to much larger values ($\sim 100 \text{\AA}$) at higher fluorine coverages.

3. Magnitude of the yields. Very large etch yields (~ 20-25 Si atoms/1 keV Ar^+ ion)[7,8] have been reported for the silicon-fluorine system in the presence of an adequate flux of fluorine from the gas phase. Such large yields are not expected from a room temperature surface on the basis of known physical sputtering phenomena. Large yields are observed in the sputtering of frozen gases[32] where the species are bound very weakly to the surface.

4. Dependence of yield on projectile mass. In situations where physical sputtering is clearly involved, the sputtering yield for He^+ is usually at least a factor of 10 less than that of Ar^+ in the energy range considered in this discussion (~ 1 keV). However, it has been found that the etch yield of 1 keV He^+ is about 40 percent of the yield of 1 keV Ar^+ on fluorinated silicon.[8] This observation is suggestive of phenomena other than physical sputtering.

5. Angular distribution of product species. In physical sputtering the yield increases as the angle of incidence is increased from zero (normal incidence) reaching a maximum in the vicinity of 60°, after which the yield decreases rapidly because glancing incidence ions are reflected with relatively little loss of kinetic energy to the surface. This trend has not been seen[19,33]

in systems involving volatile etch products suggesting again that phenomena other than physical sputtering are important in ion-assisted chemical etching.

6. Sputtering yields in the presence of an active gas. One rationale for the large etch yields observed in ion-assisted chemical etching is a chemically induced decrease in the surface binding energy.[18,19] In situations where a substrate is exposed to an energetic ion beam and a flux of reactive gas chosen so that only involatile products are formed, the etch rate of the substrate is almost never increased by the presence of the active gas.[5] In some of the chemical systems studied, the chemisorption of the active gas tends to decrease the surface binding energy.[5] These include O_2 on W, Mo, Cr, V and Cl_2 on Cu and Al. However, the increase in the sputter etch yield anticipated from the decrease in the surface binding energy is not observed, but in fact the yield is decreased. This is probably a result of the dilution of the substrate atoms in the outermost atomic layers caused by the presence of the chemisorbed gas. Thus it is not clear that the sputter etch yield of silicon will be increased by halogenation of the silicon surface.

7. Ion-induced detrapping. Very recently McFeely,[34] using X-ray photoelectron spectroscopy with synchrotron radiation, has observed some SiF_4 in a fluorinated silicon surface. Presumably this SiF_4 has been formed below the top surface where it is physically trapped. The release of this trapped SiF_4 could be accomplished by ion bombardment and mechanistically is distinct from both physical sputtering and chemical sputtering. This mechanism could be called "detrapping" and was independently suggested to the authors by McFeely[34] and Sanders.[35] If this were a dominant mechanism, both the very low value of E_b and the E^{-2} high energy dependence of the product energy distribution could be interpreted. However, preliminary estimates of the amount of SiF_4 present in the near surface region do not allow the assignment of ion-assisted detrapping as a dominant mechanism.

The preceding discussion has been centered around which of the etching steps B (product formation) or C (product desorption) is directly accelerated by energetic ion bombardment. There is also the point of view of Flamm and Donnelly[36,37] who suggest that step A (gas-surface reaction step) is directly influenced by the energetic ion bombardment. This model is based on the ion bombardment breaking silicon-silicon bonds in the surface region and thus rendering the surface more susceptible to fluorination by subsequently arriving gas phase fluorine. Experiments in this laboratory involving sequential ion bombardment and fluorination[38] have indicated that damage-induced effects are sometimes present but not dominant. In the Si-XeF_2 system, effects of damage are sometimes seen but often no effects can be observed at all. The cause of this irreproducibility is currently being examined. Larger effects of damage were seen with W(111)-XeF_2[38] but even in this system the damage-enhanced chemistry does not appear to be the dominant process involved in ion-assisted etching. The product energy distributions, which strongly suggest the involvement of the collision cascade, are inconsistent with a damage-enhanced mechanism being dominant in ion-assisted etching. Finally, the fact that etch yields on pre-halogenated surfaces (in the absence of gas phase fluorine) has been found to be comparable to etch yields measured in steady state[7,12,13] is support for the *primary* effect of the ion bombardment residing in steps B and/or C. If this is the case, damage effects can still play a secondary role in ion assisted chemical etching by increasing the reaction coefficient for fluorine on silicon at very low fluorine coverages. Suppose that the reaction coefficients for fluorine on damaged and undamaged silicon were equal at high fluorine coverages but the reaction coefficient on damaged silicon was larger at low coverages. The authors point of view would have chemical sputtering responsible for the decreased fluorine coverage but the effect of the reduced coverage could be magnified by surface damage.

SUMMARY AND CONCLUSIONS

A discussion has been presented summarizing the factors which must be taken into consideration in determining the mechanistic role of energetic ion bombardment in the ion-assisted etching process. The present authors conclude that, when these factors are considered with the experimental data presently available for the silicon-fluorine system, the primary role of energetic ion bombardment is to accelerate the formation of volatile products from the fluorinated silicon surface layer. However, this conclusion is not easily extended to other chemical systems; a fact which is emphasized by the observation that no ion enhancement is observed in the $Al-Cl_2$ system[9-11] (a retardation is seen at low Cl_2 fluxes) and ion bombardment of $Cu(111)$ exposed to Cl_2 at temperatures above 200°C results in a large decrease in the etch rate.[35]

Finally, it is important to discuss briefly the extent to which terminology and/or semantics may introduce confusion or misunderstanding into this discussion on basic mechanisms. One might take the point of view that any product ejection process originating in the collision cascade is associated with physical sputtering, or, that the product mix and product energy distribution evolved from chemical sputtering or detrapping be characteristic of a thermal process with a characteristic temperature close to that of the surface. We believe that these simplifications may tend to obscure important mechanistic insight which we anticipate will be clarified by future basic studies like those discussed above.

REFERENCES

1. C. J. Mogab, in "VLSI Technology," S. M. Sze, editor, McGraw-Hill, New York (1983).
2. J. A. Mucha and D. W. Hess, *Amer. Chem. Soc. Symp. Ser.* **219**, 215 (1983).
3. J. W. Coburn and H. F. Winters, *Ann. Rev. Mater. Sci.* **13**, 91 (1983).
4. R. A. Gottscho and T. A. Miller, *Pure and Appl. Chem.* **56**, 189 (1984).
5. H. F. Winters, J. W. Coburn and T. J. Chuang, *J. Vac. Sci. Technol.* **B1**, 469 (1983).
6. J. W. Coburn and H. F. Winters, *J. Appl. Phys.* **50**, 3189 (1979).
7. Y.-Y. Tu, T. J. Chuang and H. F. Winters, *Phys. Rev.* **B23**, 823 (1981).
8. U. Gerlach-Meyer, J. W. Coburn and E. Kay, *Surf. Sci.* **103**, 177 (1981).

9. H. F. Winters, to be published.
10. D. L. Smith and R. H. Bruce, *J. Electrochem. Soc.* **129**, 2045 (1982).
11. D. L. Smith and P. G. Saviano, *J. Vac. Sci. Technol.* **21**, 768 (1982).
12. E.-A. Knabbe, J. W. Coburn and E. Kay, *Surf. Sci.* **12**, 427 (1982).
13. J. W. Coburn, in "Applications of Piezoelectric Quartz Crystal Microbalances," C. Lu and A. W. Czanderna, editors, Elsevier, Amsterdam (1984), p. 221.
14. H. F. Winters, *J. Appl. Phys.* **49**, 5165 (1978).
15. T. J. Chuang, *J. Appl. Phys.* **51**, 2614 (1980).
16. J. F. Norar, F. R. McFeely, N. D. Shinn, G. Landgren and F. J. Himpsel, *Appl. Phys. Lett.* **45**, 174 (1984).
17. M. A. Loudiana, A. Schmid, J. T. Dickinson and E. J. Ashley, to be published.
18. J. L. Mauer, J. S. Logan, L. B. Zielinski and G. C. Schwartz, *J. Vac. Sci. Technol.* **15**, 1734 (1978).
19. T. M. Mauer and R. A. Barker, *J. Vac. Sci. Technol.* **21**, 757 (1982).
20. H. F. Winters and F. A. Houle, *J. Appl. Phys.* **54**, 1218 (1983).
21. H. F. Winters, *J. Vac. Sci. Technol.* **B1**, 927 (1983).
22. R. A. Haring, A. Haring, F. W. Saris and A. E. deVries, *Appl. Phys. Lett.* **41**, 174 (1982).
23. M. J. Vasile, *J. Appl. Phys.* **54**, 6697 (1983).
24. M. J. Vasile and F. A. Stevie, *J. Appl. Phys.* **53**, 3799 (1982).
25. D. L. Perry and J. L. Margrave, *J. Chem. Educ.* **53**, 696 (1976).
26. D. L. Flamm, D. E. Ibbotson, J. A. Mucha and V. M. Donnelly, *Solid State Technol.* **26**(4), 117 (1983).
27. R. A. Barker, T. M. Mayer and W. C. Pearson, *J. Vac. Sci. Technol.* **B1**, 37 (1983).
28. H. F. Winters, unpublished data.
29. F. H. M. Sanders, A. W. Kolfschoten, J. Dieleman, R. A. Haring, A. Haring and A. E. deVries, *J. Vac. Sci. Technol.* **A2**, 487 (1984).
30. A. W. Kolfschoten, R. A. Haring, A. Haring and A. E. deVries, *J. Appl. Phys.* **55**, 3813 (1984).
31. J. Dieleman and F. H. M. Sanders, *Solid State Technol.* **27**(4), 191 (1984).
32. S. K. Erents and G. M. McCracken, *J. Appl. Phys.* **44**, 3139 (1973).
33. H. Okano and Y. Horiike, *Jpn. J. Appl. Phys.* **20**, 2429 (1981).
34. F. R. McFeely, private communication.
35. F. H. M. Sanders, private communication.
36. V. M. Donnelly and D. L. Flamm, *Solid State Technol.* **24**(4), 161 (1981).
37. D. L. Flamm and V. M. Donnelly, *Plasma Chem. Plasma Process* **1**, 317 (1981).
38. H. F. Winters, to be published.

DAMAGE EFFECTS IN REACTIVE ION ETCHING

S. J. Fonash
Engineering Science Program
The Pennsylvania State University
University Park, PA 16802

ABSTRACT

Reactive ion etching can cause extensive, electrically active damage. This damage can be categorized as (i) impurity and etching-ion implantation, (ii) residue and film formation, and (iii) intrinsic bonding damage. Intrinsic bonding damage is the most difficult to deal with since it appears inherent to dry etching processes which employ directed ion bombardment.

The presence of this bonding damage can be established with electron spin resonance (ESR) which demonstrates that RIE produces the broad, isotropic paramagnetic resonance signal which characterizes trivalently bonded Si defects. In addition, the presence of this damage can be made readily apparent by simply fabricating metal-semiconductor contacts on reactive ion etched surfaces. For silicon, such contacts, using non-reactive metals, yield barrier heights which are reduced below expected values for n-type material and increased above expected values for p-type material. This result indicates the presence of positive charge in reactive ion etched Si and establishes that the effect is not one of simply increasing the surface recombination speed or one of simply introducing the series resistance of a surface residue layer.

INTRODUCTION

Dry etching processing is crucial for obtaining the fine line definition, highly directional etching, and selectivity required in VLSI fabrication. The dry etching processes that can give these desirable features are plasma etching (PE) under certain operating conditions, reactive ion etching (RIE), reactive ion beam etching (RIBE), magnetron reactive ion etching (MRIE), and chemically assisted ion beam etching (CAIBE).[1] These techniques involve various geometries and various electrode configurations but all are based on using directed ion bombardment in the etching process.

Although these techniques are essential for obtaining the etching control needed in VLSI, it is found, unfortunately, that they can damage semiconductors and dielectrics. This damage may be categorized as (i) impurity and etching-ion implantation,[2-4] (ii) residue and film formation,[2-5] and (iii) intrinsic lattice or bonding damage.[2,4,6-9] The first two types of dry etching damage on this list can, in principle, be controlled by use of carefully designed equipment and by the avoidance of chemistries that result in film or residue production. The last type of damage - intrinsic lattice damage - is the most insidious since it is a consequence of the directional ion bombardment required for anisotropic dry etching.

When present, surface film and residue formation can lead to detrimental effects such as high contact resistance. Further, impurity implantation and lattice defects can result in detrimental effects such as neutral traps and positive charge in SiO_2.[10,11] They can produce modified electrical properties for SiO_2/Si and metal/Si interfaces.[10-15] In addition, the presence of damage at a dry etched silicon surface can affect the growth of silicides on these surfaces[16] and, recently, it has been shown that sputtering damage, similar to the bonding damage caused by the dry etching used in VLSI, can enhance Aℓ contact penetration into silicon.[17]

In this paper we will focus on one type of dry etching damage - what we have called the intrinsic lattice or bonding damage - since it appears to be inherent to directional dry etching processes. As we have noted, the other classes of damage, residue formation and impurity implantation, need not be inherent. Our discussions will also focus on one type of dry etching; viz, reactive ion etching. However, the observations and conclusions obtained based on this study of RIE lattice damage are transferable to the other dry etching techniques using directed ion bombardment. To establish that RIE does damage VLSI materials, we will explore the properties of Si surfaces, such as that seen in Fig. 1, which have been subjected to reactive ion etching.

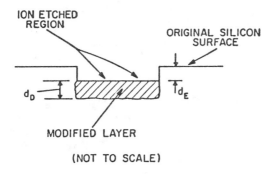

Fig. 1. Silicon surface that has been subjected to reactive ion etching (RIE).

All the silicon samples to be discussed here were etched in an RIE system with a water-cooled, stainless-steel cathode and a perforated anode held at ground. Pressure was kept in the 20-100 μm range and the rf power was varied between 100-600 watts. Pressure, power, and peak cathode voltage (approximately equal to the energy of the ions impinging on the Si) were monitored in all etching runs.

DETECTION OF RIE DAMAGE

A. Spreading Resistance Measurements

A straight forward approach to determining if RIE modifies the electrical properties at the surface of etched silicon is to evaluate etched material using spreading resistance measurements. The resulting measurement yields a resistance which is the series combination of the metal probe/silicon surface contact resistance and the constriction resistance in the silicon beneath the probe.

Figure 2 shows spreading resistance data for Si subjected to CCl_4 RIE. This spreading resistance measurement was undertaken using the single probe technique and, to minimize penetration, a probe loading of only 5 grams was used which resulted in a probe "footprint" of only $\sim 5\mu$. The figure has two typical spreading resistance maps. One has been obtained by stepping the probe from unetched to the RIE region on n-type Si and the other has been obtained by stepping the probe from unetched to the RIE region on p-type Si. As may be noted from the figure, the spreading resistance was found to be higher in RIE regions on p-type Si and lower in RIE regions on n-type Si. The use or omission of an HF dip procedure prior to the measurements was found to have no significant effect on this trend. This HF procedure consisted of a 5 sec. dip in 1:20 $HF:H_2O$ just prior to the measurement. It was explored to determine if different oxidation rates, between RIE etched and non-RIE etched regions, could cause the results seen in Fig. 2.

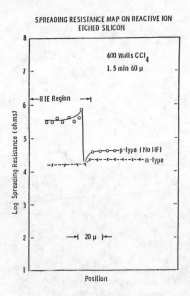

Fig. 2. Spreading resistance data for Si samples subjected to CCl_4 RIE. Both n and p materials were Cz-grown 1-10 Ω-cm.

The data of Fig. 2 cannot be explained in terms of insulating surface residues or films since these would introduce series resistance for both n and p-type Si. These data also cannot be explained in terms of an increased surface recombination speed due to damage. If increased surface recombination were the only phenomenon present, then contacts to n and p-Si would improve in RIE exposed regions. Unfortunately, because of the nature of the measurement, it cannot be determined from Fig. 2 if it is the contact characteristics or the constriction resistance which is being modified in the RIE regions. However, by going to larger, deposited metal/silicon contacts, constriction resistance can be suppressed and the current-voltage (I-V) measurements can focus on the electrical characteristics of the metal/silicon interface itself. This is the approach of the next section.

B. Metal/Si I-V Characteristics

The current-voltage (I-V) characteristics of metal-semiconductor contacts have been very successfully employed as a probe of surface damage in RIE and IBE etched Si.[3,4,7-9,12-16,18] In general, examination of these characteristics has shown that dry etching of Si using directed ion bombardment (IBE and RIE) causes apparent barrier lowering, when compared to controls on chemically etched n-type Si, for contacts between non-reactive metals and dry etched n-type Si. Further, it causes barrier enhancement, when compared to controls on chemically etched p-type Si, for contacts between non-reactive metals and dry etched p-type Si. This phenomenon has been observed in the case of IBE for etching with H, Ne, Ar, and Kr[3,4,7-9] and it has been observed in the case of RIE for etching with NF_3[13,18] and for etching with CF_4, CCl_4, or CCl_4 + He.[14]

We note that this barrier lowering which occurs as a result of directed ion bombardment for n-type silicon and barrier raising which occurs as a result of directed ion bombardment for p-type silicon is observed for non-reactive metals such as Au. That is, for non-reactive metals it is clear that damage incurred (as a result of ion bombardment) creates donor states in a surface layer of the etched Si. Equivalently, the damage can be thought of as pinning the Fermi level above the middle of the gap at the etched surface. This easily discerned barrier modification trend becomes somewhat clouded if silicide-forming metals are used since barrier heights can change if, and when, the metal and silicon react. Further, silicide formation can consume the damaged layer present in the Si due to dry etching. These points are discussed elsewhere in this conference[19] and in ref. 16.

This barrier shifting phenomenon for a non-reactive metal (such as Au) is shown in Figures 3 and 4 for CCl_4 or CCl_4 + He RIE etching of Si. Figure 3 shows the current-voltage (I-V) characteristics for Au/p-Si contacts fabricated on Si subjected to various rf power levels in CCl_4 + He RIE. Ohmic-like (very low barrier height) characteristics are seen, as expected, for the contact on the chemically-etched only (without RIE) p-Si. However, a barrier is seen to be present for Au contacts on RIE etched p-Si and this barrier is more pronounced with increasing RIE rf power for a constant pressure, constant etching time and constant gas flow rate.

The corresponding effect on n-type Si is shown for CCl_4 RIE treated n-Si in Figure 4. Here the data are presented for various

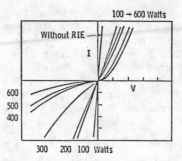

Fig. 3. Current-voltage characteristics for Au/p-Si contacts fabricated on silicon which has been subjected to CCl_4 + He RIE with various rf power levels. Samples were etched for 1-1/2 minutes.

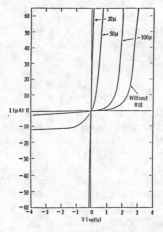

Fig. 4. Current-voltage charactersitics for Au/n-Si contacts fabricated on silicon which has been subjected to CCl_4 RIE with various gas pressure levels. Samples were etched for 1-1/2 minutes at 300 watts.

gas pressures. A high barrier height, Schottky characteristic is obtained, as expected, for the Au contact on the chemically-etched only (without RIE) n-Si. However, this barrier is seen to degrade on RIE etched n-Si and the degradation is more pronounced with decreasing RIE gas pressure for a constant rf power level, constant etching time, and constant gas flow rate.

Figures 3 and 4 make the point that CCl_4 and CCl_4 + He etching both damage Si. This damage is at the immediate surface of the Si since the Au/Si contact barrier height has been affected by the use of RIE. Figures 3 and 4 also make the point that RIE exposure does not simply introduce damage that increases the surface recombination speed. If this were the case, the barrier height of Au/Si contacts would appear to be reduced after RIE for both n-type and p-type silicon. The barrier shifts observed are consistent with the model that RIE damage, as well as IBE damage, introduces localized gap states (donor states), which contain positive charge for Au/Si contacts, into a modified surface layer.[3,4,7,8,12-16,18] The data of Figs. 3 and 4 cannot be interpreted in terms of an insulating surface film or residue left by the RIE. If such a film or residue were dominating, it would appear as a series resistance and increase contact resistance on n and p-type Si. To avoid any surface films, all the samples of Figs. 3 and 4 were given a mild HF etch prior to the Au deposition. Figs. 3 and 4 show that the damage caused by low energy ion bombardment increases with increasing rf power level and also increases with decreasing RIE gas pressure. This dependence of the damage on increasing rf power level and on decreasing gas pressure can be explained by the data of Figure 5: these show that the cathode voltage, a measure of the energy ions pick up while crossing the sheath above the silicon samples, increases with rf power and also increases with decreasing pressure. The data of Figure 5 were measured in an Anelva etching station.

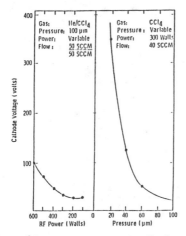

Fig. 5. Cathode voltage, a measure of the energy ions acquire while crossing the sheath above the silicon being etched, as a function of rf power or gas pressure.

DAMAGE MODEL

Although surface films and residues can remain on a silicon surface after RIE, this type of damage effect is clearly not the cause of the behavior seen in Figs. 3 and 4. However, the donor levels (or, equivalently, the Fermi level pinning) causing the behavior seen in Figs. 3 and 4 could be due to implanted etching species or implanted impurities. The former can be ruled out since it has been shown that the same barrier lowering on n-Si and barrier raising on p-Si can be caused by a number of non-reactive (Ne, Ar, Kr) as well as reactive (NF_3, CF_4, CCl_4) etching species. The latter can be ruled out on the basis of the following observation: it has been found that a mass analyzed 3 keV Ar^+ beam causes the same damage in silicon; i.e., exposure to the beam lowers the barrier for n-type Au/Si contacts and raises the barrier for p-type Au/Si contacts.[20] Hence, the only remaining possible cause of the positive charge residing in the damaged Si surface layer is bonding (or lattice) damage.

In this section we report on efforts to establish that Si bonding damage is present in that modified layer produced by RIE. To make that determination we used electron spin resonance (ESR) measurements. We have previously shown, using ESR, that silicon bonding damage (in the form of a type of trivalently bonded silicon defect) is present in IBE etched silicon.[4-7] In that work a correlation was established between the I-V behavior of Au/Si contacts, ESR detected bonding damage, and the energy of the impacting ions.[7] However, due to sensitivity limitations, the Si bonding damage incurred in IBE was only traced down to impacting-ion energies of 700 eV. To increase the signal strength for this study, we placed five identically prepared ESR samples simultaneously in the ESR cavity. Figure 6 gives the ESR spectrum, prior to RIE exposure, for the magnetic field H_o perpendicular to the surface of these samples and also for H_o parallel to the surface of the samples. The signature seen in these spectra is that of the

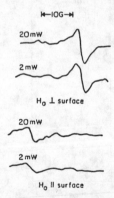

Fig. 6. Electron spin resonance signal for five Si samples prior to RIE exposure. The signature of the so-called P_b defect center is seen.

so-called P_b defect center.[21,22] This center has been established as a trivalently bonded Si defect existing at an SiO_2/Si interface. It is known that Si that has been freshly etched in acid displays no P_b defects; however, they appear as a nascent oxide begins to grow and are also present in thermally grown oxides.[22] As may be seen, the structure of these P_b spectra is definitely anisotropic (i.e., signal strength and magnetic field dependence is a function of magnetic field orientation) which is a result of there being a preferred orientation for the disrupted bond at the SiO_2/Si interface.[21,22]

Figure 7 shows the ESR signature of these same five specimens after RIE exposure. The spectrum is seen to be almost isotropic now and essentially the same as the spectrum obtained previously for ion beam etched silicon.[4,7] This spectrum seen in Fig. 7 is the signature of trivalently bond silicon produced by damage. There is also some indication in Figure 7 of P_b resonances superimposed on this isotropic damage signal which is to be expected, since the ESR measurements were done several days after the RIE exposure. Comparison of Figures 6 and 7

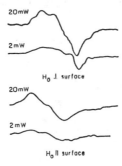

Fig. 7. Electron spin resonance signal for the same five Si samples of Fig. 6 after RIE exposure. The signature of the isotropic trivalently bonded Si lattice damage defect is seen.

makes the point that reactive ion etching does result in silicon bonding damage. This bonding damage is the same as that previously seen for IBE-etched Si; i.e., it is a type of trivalently bonded silicon defect.

The trivalently bonded silicon defect site giving rise to the signature seen in Fig. 7 is depicted schematically in Fig. 8a. As seen in Fig. 8a, to contribute to the ESR signal this defect must have an unpaired electron. That is, when that unpaired electron is present and the defect is contributing to the ESR signal, the site must be neutral. As also seen in Fig. 8 these sites can be positively charged (donors) when the unpaired electron leaves and these sites can be negatively charged (acceptors) when an additional electron is captured. Hence, the bonding defect detected by ESR is amphoteric; i.e., it can be a donor or acceptor.

The model that emerges is that the ion bombardment integral to RIE causes trivalently bonded silicon defect sites in a surface layer of etched Si. Those sites that are neutral (i.e., retain the unparied electron) show up in the ESR measurements. A certain frac-

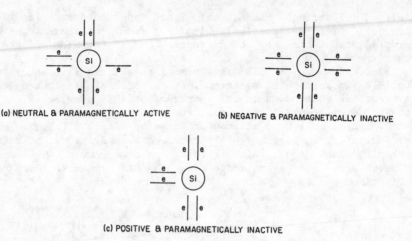

Fig. 8. Schematic of the Si trivalently bonded defect site. This site is seen to be amphoteric.

tion of these sites could be negative (those that have gained another electron) and a certain fraction could be positive (those that have lost the unpaired electron). Neither the negatively charged nor the positively charged sites are detected by ESR. Phenomenologically, the data suggest that the number of positively charged sites must be larger than the number of negatively charged sites in the damaged surface layer caused by RIE in order to obtain the barrier shifts observed. Why donor sites dominate over acceptor sites remains an open question at this time.

To further explore the state of RIE exposed or IBE exposed silicon surfaces we turned to photo-conductive resonance measurements.[23] This technique uses light to drive the populations of carriers, near the surface of silicon, above their thermodynamic equilibrium values. The recombination of these carriers is retarded, if it takes place through a paramagnetically active recombination center, by the presence of a magnetic field since the magnetic field aligns electron spins in the material. However, this recombination level will be Zeeman split due to the presence of the magnetic field and new spin orientations can be introduced if electrons are mixed between these two Zeeman levels. Microwave photons will provide such mixing. Consequently, when the microwave photon energy equals the Zeeman splitting of the paramagnetically active recombination center, recombination will be enhanced and photoconductivity will decrease. Photo-conductive resonance (PCR) directly measures the resulting change in the Q of an ESR cavity which occurs when recombination becomes more efficient due to microwave-caused mixing.

Figure 9 is the PCR signal for one of the five samples used to generate the ESR data of Figures 6 and 7. Figure 9a is the PCR signal prior to RIE and Figure 9b is the PCR signal after RIE. It

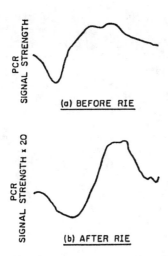

Fig. 9. Photoconductive resonance signal for one of the five samples of Figs. 6 and 7. Fig. 9a is the PCR signal before RIE; Fig. 9b is the PCR signal after RIE.

is clear that RIE greatly diminishes the PCR signal; it was found that IBE extinguishes this signal for IBE exposed samples. The chemical defect (the recombination center) giving rise to the PCR signal seen in Figure 9a is not established at this time.[24] However, it has been suggested that it is related to hydroxyl or hydrogen defects in nascent SiO_2. Since the measurements of Figure 9a were taken several days after acid etching and those of Figure 9b were taken a number of days after RIE exposure, the data of this figure indicate that the native SiO_2 layer is not growing with the same defect structure on RIE (or IBE) etched Si that it has on acid-etched Si surfaces. The result is the differing surface recombination behavior seen in Figure 9.

DAMAGE CONTROL APPROACHES

Residues and surface film RIE damage can be eliminated by going to etching chemistries that do not invoke the formation of such coatings. Further, it has been demonstrated that impurity incorporation can be suppressed by covering etch chamber walls with protective coatings.[3] Current approaches to coping with the fundamental lattice damage problem have included wet etching, annealing, consumption (as with silicide formation), and passivation with low energy hydrogen implants. Clearly wet etching certainly will not be acceptable in future VLSI device fabrication and elevated temperature furnace annealing also may not be acceptable. It has been suggested that an acceptable approach to controlling all three forms of RIE damage (residues, impurity and etching-ion implantation, and the fundamental problem of bonding damage) may be to follow anisotropic etching, done to depths that cause no damage to layers that must remain intact, with non-damaging, isotropic dry chemical etching.[2] This type of dry etching does not involve directed ion bombardment.

We turn to Fig. 10 to demonstrate the temperatures needed to anneal out the fundamental bonding (lattice) damage created by RIE. Fig. 10 shows the I-V characteristics of several Au/Si contacts fabricated on p-type silicon. In each case the silicon has been reactively ion etched using CCl_4 with a power of 600 W, a pressure of 60 μ, and an etching time of 1-1/2 min. For the sample yielding Curve a the Au contact was deposited after RIE exposure. However, the sample that yielded Curve b was annealed after RIE for 1 hour at 500°C in Ar and the sample that yielded Curve C was annealed after RIE for 1 hour at 600°C in Ar. The Au contacts for samples b and c were then deposited after the anneals.

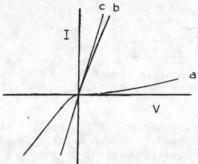

Fig. 10. Current-voltage characteristics of Au/p-Si contacts fabricated on Si subjected to CCl_4 600 W, 60 μ (1-1/2 min) RIE. For curve a the Au was deposited after RIE, for curve b the sample was annealed for 1 hr. at 500°C in Ar after RIE and then the Au was deposited, and for curve c the sample was annealed for 1 hr. at 600°C in Ar after RIE and then Au was deposited. Curve c also results if an 800°C anneal is used.

The data of Figure 10 show that the surface is only partially restored after the 500°C/1 hour anneal; a 600°C/1 hour anneal is required for full recovery. Although these temperatures are lower than those needed to anneal out IBE damage,[4] the use of 600°C anneals may become a matter of concern for future devices and for future fabrication strategies.

Low temperature silicide formation can be used to reduce the effects of the intrinsic bonding damage caused by RIE; but, as discussed in detail elsewhere,[16] low temperature silicide formation does not completely remove damage effects. To be specific, low temperature silicide formation is found to reduce the damage-caused barrier on p-type Si so that recovery to the expected ohmic-like I-V behavior (for high work function metals) is observed. However, recovery of the barrier on n-Si to the high value expected (for high work function metals) is not observed.[16,19]

We briefly touch here on the use of low energy hydrogen ion implants for the passivation of RIE damage since we have discussed the subject elsewhere in detail.[8] It has been established that the modified electrical properties detected by thermally deposited Au/Si contacts arise from a layer that has been found to contain silicon bonding defects. Hydrogen might be expected to exercise its well-known ability to tie-up dangling Si bonds if introduced into these damaged layers. Further, if these dangling bonds are related to the anomalous electrical properties, hydrogen passivation should restore RIE etched Si to a state where Au/Si contacts exhibit normal (i.e., the same as seen on chemically etched Si) behavior.

Hydrogen implants are found to be able to do just what is expected; i.e., they can passivate the modified layer produced by RIE as seen in Figure 11. As discussed elsewhere,[4] low energy hydrogen implants can also produce an insulating film on the surface of the Si, as may be noted from Figure 11, which must be removed. It has been established that removal of this film is not the cause of surface restoration;[8,9] i.e., it is the tying up of dangling Si bonds by hydrogen which causes passivation.

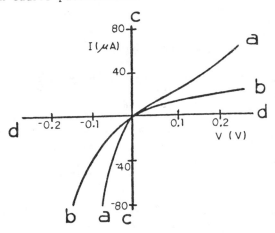

Fig. 11. Hydrogen passivation of RIE damage in p-Si. Samples a and b have no H^+ passivation. Samples c and d have been subjected to a 5 min. 0.4 keV H^+ exposure. Samples a and c had HF etches prior to Au deposition.

CONCLUSIONS

Spreading reistance measurements have been able to detect the presence of a modified surface layer for CCl_4 or CCl_4 + He reactive ion etched silicon. Thermally deposited Au/Si contacts to these same modified layers display anomalous I-V characteristics similar to those previously observed for Au/Si contacts to NF_3 RIE exposed Si and similar to those previously observed for Au/Si contacts to H, Ne, Ar, or Kr IBE exposed Si. The anomalous I-V behavior of RIE etched layers cannot be attributed to insulating surface residues left by RIE since they would always lead to increased contact resistance. It does not appear that the anomalous I-V behavior can be attributed to implanted impurities either since the same type of characteristics are obtained for contacts deposited on Si surfaces etched with a mass-analyzed 3 keV Ar beam.

The shift in the barrier height for the Au/Si contacts on surfaces RIE etched with CCl_4 or CCl_4 + He indicates that the modified surface layer has localized states, which for non-reactive metal/Si contacts, contain positive charge (i.e., there are donor levels present). This same observation, that the modified layer caused by etching must contain positive charge for non-reactive metal/Si contacts, has been made earlier in the investigation of NF_3 RIE etching, in the investigations of H, Ne, Ar, and Kr IBE etching, and in the investigation of the mass-analyzed 3 keV beam exposure. Studies of the degree of Au/Si barrier shift on an RIE etched surface as a function of power or pressure in this investigation demonstrated that the degree of barrier shift is controlled by the energy of the etching ions as they impact the silicon surface.

A search for silicon bonding defects in this modified layer produced by CCl_4 or CCl_4 + He RIE established that a trivalently bonded silicon defect is created by RIE exposure. This is the same defect found to be present, in an earlier study, as a result of ion beam etching of Si.

ACKNOWLEDGEMENTS

This work was supported in part by the National Science Foundation under Grant #ECS-8305646 and in part by the Department of the Army under Contact #DAAK 20-84-K-0456.

The author wishes to acknowledge A. Climent, R. Davis, P. Lester, R. Singh, and J.-S. Wang who worked on this investigation. He wishes to thank E. Poindexter and P. Caplan for the ESR and PCR measurements and A. Rohatgi for some of the sample preparation. Thanks are also due to M. Chin for the plasma measurements.

REFERENCES

1. S. J. Fonash, Solid State Technology, in press.
2. J. Dieleman and F. H. M. Sanders, Solid State Technol. 27(4), 191 (1984).
3. S. W. Pang, Solid State Technol. 27(4), 249 (1984).
4. R. Singh, S. J. Fonash, S. Ashok, P. J. Caplan, J. Shappirio, M. Hage-Ali, and J. P. Ponpon, J. Vac. Sci. Technol. A1, 334 (1983).
5. J. S. Chang, Solid State Technol., 27(4), 214 (1984).
6. P. Caplan, E. Poindexter, A. Rohatgi, P. Lester, R. Singh, and S. J. Fonash, J. Electrochem. Soc., submitted for publication.
7. R. Singh, S. J. Fonash, P. J. Caplan, and E. H. Poindexter, Appl. Phys. Lett., 43, 502 (1983).
8. J.-S. Wang, S. J. Fonash, S. Ashok, Electron Device Lett., EDL-4, 432 (1983).
9. R. J. Davis, R. Singh, S. J. Fonash, P. J. Caplan, and E. H. Poindexter, Symposium Proceedings, Materials Research Society Meeting, Boston 1983, in press.
10. L. M. Ephrath and D. J. DiMaria, Solid State Technol. 24, 183 (1981).

11. D. J. DiMaria, L. M. Ephrath, and D. R. Young, J. Appl. Phys. 50, 4015 (1979).
12. S. J. Fonash, S. Ashok, and R. Singh, Thin Solid Films 90, 231 (1982).
13. S. Ashok, T. P. Chow, and B. J. Baliga, Appl. Phys. Lett., 42, 687 (1983).
14. A. Rohatgi, M. R. Chin, P. Rai-Choudhury, P. Lester, R. Singh, and S. J. Fonash, Ext. Abs. 320, 164th Meeting of Electrochem. Soc., Oct. 1983.
15. S. W. Pang, D. D. Rathman, D. J. Silversmith, R. W. Mountain, and P. D. DeGraff, J. Appl. Phys. 54, 3272 (1983).
16. A. Climent and S. J. Fonash, J. Appl. Phys., 55 (1984).
17. L. J. Brillson, M. L. Slade, A. D. Katnani, M. Kelly and G. Margaritondo, Appl. Phys. Lett. 44, 110 (1984).
18. T. P. Chow, S. Ashok, B. J. Baliga, and W. Katz, J. Electrochem. Soc. 131(1), 156 (1984).
19. C. Mu, A. Climent, and S. J. Fonash, this conference. (See also Ref. 16).
20. S. Ashok and A. Mogro-Campero, IEEE Electron Dev. Lett., EDL-5, 48 (1984).
21. P. J. Caplan, E. H. Poindexter, B. E. Deal, and R. R. Razouk, J. Appl. Phys. 50, 5847 (1979).
22. P. J. Caplan, E. H. Poindexter, and S. R. Morrison, J. Appl. Phys. 53, 541 (1982).
23. J. Shiota, N. Mujanota, and J. Nishizava, J. Appl. Phys. 45, 2556 (1977).
24. G. C. Roberts, M. C. Petty, P. J. Caplan, and E. H. Poindexter in press.

SIMULATION AND ANALYSIS OF ANOMALOUS PLASMA ETCHING

S.Kawazu, T.Nishioka and H.Koyama

LSI R&D Laboratory, Mitsubishi Electric Corp.
4-1 Mizuhara Itami Hyogo, 664 Japan

ABSTRACT

This paper deals with a SIMS analysis and a physical simulation of an anomalous reactive ion etching of poly silicon for understanding the physical environments of the plasma processing. The symmetric perturbation of the plasma cylinder affected the surface migration of iron compounds resulting in symmetrically localized anomalous etching of poly silicon.

INTRODUCTION

Plasma assisted etching is one of the key elements of the VLSI manufacturing technology. When a semiconductor device feature reaches a 1 μm dimension, reactive ion etching (RIE) becomes essential for the fine pattern definition. However, the plasma process depends on a multitude of parameters and investigations into plasma technology as applied to semiconductor device fabrication so far are non-quantitative or semi-quantitative in nature.

Correspondingly, a better understanding of the physics and chemistry of plasma etching is strongly required to solve commonly observed reproducibility and reliability problems associated with plasma processes. The present paper deals with an analysis of an anomalous reactive ion etching of poly silicon and a physical simulation of the phenomenon for the better understanding of the physical environments of the plasma processes.

EXPERIMENT

The present work was done in a parallel plate reactor used in the RIE mode. The 13.56 MHz rf-power was applied to the lower electrode. Partially isotropic etching of poly-silicon in $C_2Cl_2F_4$ gas was carried out at 14 Pa. The diameter of the upper and lower electrodes were 17 cm and 14 cm, respectively. The distance between the electrodes was 4 cm.

EXPERIMENTAL RESULTS

The anomalously etched pattern which is beautifully symmetrical in shape is shown in Fig.1. Surface undulation formed after the reactive ion etching of polysilicon represents the anomalous pattern. Surface residues looked like dendrites are observed in an SEM (Scanning Electron Microscope) picture as shown in Fig.2. The distribution of these residues seem to be related to the present symmetrical pattern, that is the residues distribute preferentially at the "petal" site in the symmetrical pattern.

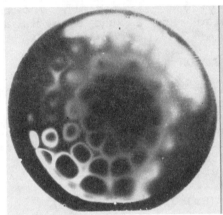
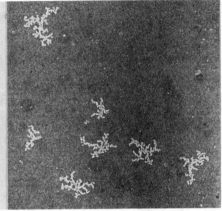

Fig.1.
Anomalous symmetric pattern appeared after the reactive ion etching of poly-silicon.

Fig.2.
SEM (Scanning Electron Microscope) picture of surface residues looked like dendrite.

High magnification SEM pictures indicate that the residues act as masks against the RIE. The surface of the residues is flat and the thickness is about 35 nm.

From the SIMS (Secondary Ion Mass Spectroscopy) analysis it is found that the material of the residues are mostly light elements as well as iron. The volatility of the residues is limited, and the materials may form compopunds such as FeF_2, Fe_2O_3, or $FeCl_2$ of which vapor pressures are much lower than that of poly-silicon.

Nucleus generation in an evaporated thin film has well been studied. It is concluded that some kinds of surface defects such as Si oxides, surface contamination and adsorbed gaseous molecules act as stepping-stones for aggregation of evaporated particles and also these defects work as nucleation sites if the probability of states exceed a certain threshold level. Stable clusters grow surrounding these activated sites.

Similar phenomenon can be expected in the present case directly after the start of plasma etching of poly-Si by $C_2Cl_2F_4$ gas. The shapes of the residues as shown in Fig.2 suggest the diffusion limited aggregation (DLA) of the Fe compounds: calculated shapes of residues based on the DLA are apparently similar to those observed as in Fig.2. It is clear, therefore, that the residues are created by the fast lateral surface migration of particles. Nucleus for the Fe cluster formation should be activated when the plasma pressure exceeds some critical point. The aggregation of Fe compound particles is affected by the plasma pressure distribution. If the intensities of plasma pressure vary symmetrically, the growth of the Fe compounds would also vary as the symmetrical pattern.

PLASMA SIMULATION (#1)

To make clear the mechanism of the anomalous pattern generation, distribution of plasma pressure is calculated next. We consider a helical magnetic field in the plasma cylinder carrying current in the reactive chamber. The plasma cylinder expressed in a cylindrical coordinates (r,φ,z) is approximated magnetohydrodynamically.(1)

Unperturbed plasma pressure p can be expressed as

$$p(r) = p_o + B_o^2/\mu_o 2\mu/\varepsilon^2 \cdot (2\mu - \varepsilon)(J_o(\varepsilon r) - 1) \qquad (1)$$

Components of $\nabla \tilde{p}$ ($\tilde{p}$ is the perturbed plasma pressure) are expressed as

$$\partial \tilde{p}/\partial r = B_o/\mu_o 2\mu \delta \varepsilon^2 (J_1' J_m' + J_1 J_m'' + [-mJ_1 J_m/(\varepsilon r)^2 + 2J_1 J_m'/(\varepsilon r)]) \cos(m\varphi)$$

$$1/r \, \partial \tilde{p}/\partial \varphi = -B_o/\mu_o \cdot 2\mu \delta \varepsilon^2 [J_1 J_m/(\varepsilon r)^2 + J_1' J_m/(\varepsilon r)] \sin(m\varphi) \qquad (2)$$

$$\partial \tilde{p}/\partial z = -B_o/\mu_o \cdot 2\mu \delta \varepsilon^2 [J_o' J_m/(\varepsilon r)] \sin(m\varphi)$$

The primes denote derivatives with respect to r, and m an order. Here, ε, δ, μ, are parameters, and μ_o is the magnetic constant of vacuum and is a constant. Terms inside the brackets [ ] are small compared to $(J_1' J_m' + J_1 J_m'')$, and can be neglected resulting in the following relation:

$$\partial \tilde{p}/\partial r = B_o/\mu_o 2\mu \delta \varepsilon^2 (J_1' J_m' + J_1 J_m'') \cos(m\varphi) \qquad (3)$$

Total plasma pressure P $(=p+\tilde{p})$ is described as

$$P = P_0 + B_0^2/\mu_0 \cdot 2\mu/\varepsilon^2 ((2\mu-\varepsilon)(J_0(\varepsilon r)-1) + \delta\varepsilon^2/B_0 J_1(\varepsilon r) J_m'(\varepsilon r) \cos(m\varphi)) \qquad (4)$$

Here P_0 and B_0 are constants.

Comparing the radial and azimuthal distribution in the anomalous fringe with the calculation it is concluded that the actual anomalous pattern is characterized by the perturbed term of the total plasma pressure.

Furthermore perturbation of plasma should vary with time at a certain frequency w. So the plasma pressure which characterizes the anomalous pattern can be written as

$$\tilde{p} = B_0 \mu_0 \delta \varepsilon^2 \cdot J_1(\varepsilon r) J_m'(\varepsilon r) \cos(m\varphi) \sin(wt) \qquad (5)$$

The pattern generation should be influenced much more by larger plasma density, and the effective plasma pressure which characterizes the anomalous pattern is finaly expressed as

$$\tilde{p}_{eff} = |\tilde{p}|_{\sin(wt)=1} \qquad (6)$$

The order m should be integers.

A distribution of intensities of the relative plasma pressure when the order m takes values as 4,6,6,8 is shown in Fig.3 and it is obvious that the simulation coincides very well with the anomalous pattern.

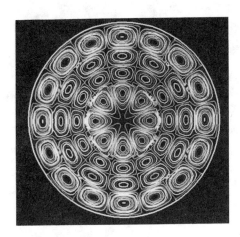

Fig.3.
Calculated contour map of plasma-pressure intensities expected in the RIE chamber.

SUMMARY

The perturbation of the plasma cylinder affected the surface migration of iron compounds resulting in symmetrically localized masks against the plasma etching. The physical environments of the plasma processing especially concerning the plasma distribution has thus been made clear, and a more sophisticated plasma etching system may be constructed through these kinds of physical considerations.

ACKNOWLEDGMENT

The authors are indebted to Dr.H.Oka and Dr.H.Nakata for their encouragement. Thanks are also due to T.Yasue for helpful discussion.

REFERENCES

1. L. S. Sovolev et al., Review of Plasma Physics (Consultants Bureau, New York - London, 1970, Translated from Russian.)

Part VI. Defects in Semiconductors

DEFECTS IN SILICON AND THEIR RELATION TO VLSI

Sokrates T. Pantelides
IBM Thomas J. Watson Research Center, Yorktown Heights, NY 10598

ABSTRACT

Three types of defects in Si are identified and discussed: The undesirable process-induced defects which must be avoided, eliminated, or passivated; the "benevolent" defects which can be used judiciously to accomplish particular objectives; and the process-controlling defects which are created by the process, but also mediate and control the process and the formation of undesirable process-induced defects. The latter, whose existence and importance is emphasized here, may prove to be the ultimate tool for process engineering. Current microscopic understanding of key VLSI-related defects is reviewed briefly. Finally, recent advances in the microscopic description of defect-mediated atomic diffusion processes are described.

INTRODUCTION

The subject of defects in silicon is very broad and varied even if we restrict our attention to defects relating to VLSI. Each class of such defects could be the subject of a full paper. In fact, other papers at this conference address specific classes of defects, e.g., defects induced by reactive-ion etching,[1] process-induced microdefects in VLSI Si wafers,[2] defects at the Si-SiO$_2$ interface,[3] etc. The purpose of this paper, on the other hand, is to give a general account of defects in Si as they relate to VLSI. It would be unreasonable, however, to attempt anything like an archival catalog of defects and their relation to VLSI. At the same time, this is not the appropriate place for a detailed description of any one defect or issue. Thus, this forum will be used to strike a number of general themes about the role of defects and our current understanding of their properties. A number of examples will be used to illustrate these themes. In the latter part of the paper, recent work by the author and his collaborators on defect-mediated atomic diffusion will be described briefly within the context of the other general issues that relate to VLSI.

THE ROLE OF DEFECTS IN VLSI SILICON PROCESSING

Silicon technology relies on a succession of processing steps that lead to the fabrication of devices. Initially, a crystal is grown and sliced into thin wafers. After that, each processing step modifies the wafer in one of three different ways: it modifies the existing crystal (e.g., by annealing, by the incorporation of impurities in an inhomogeneous way, etc.), it adds new layers of material (e.g., oxidation, metalllization, etc.), or it removes some of the material that is already present (e.g., by etching parts of previously grown oxide, etc).

Defects are endemic to all processes. From the very start, the process of growing a crystal inevitably introduces defects. These may be structural defects, such as dislocations, stacking faults, etc., or impurities such as oxygen, carbon, heavy metals, etc., either at isolated sites or in the form of complexes and precipitates. Each subsequent processing step introduces new defects which are usually generically referred to as process-induced defects.

The word "defect" implies something undesirable that spoils perfection. Indeed, when process-induced defects are mentioned, one usually thinks of undesirable defects which are responsible for reduced yields, reduced device efficiency, or reduced device lifetime. Examples of such defects are stacking faults induced by the oxidation process, dislocations that act as unwanted diffusion pipes, the so-called oxygen-related thermal donors which appear after a $350°C$ anneal and enhance the electron concentration in undesirable ways, and defects that act as recombination centers and limit the carrier lifetimes in undesirable ways. A great deal of time and effort is spent to learn how to avoid, eliminate or passivate such defects.

Nevertheless, it is also recognized that many defects have a "benevolent" role. The most obvious defects in this category are the dopant impurities such as phosphorus, arsenic, boron, etc. Indeed, these are so benevolent, that we normally do not think of them as defects. Without them, of course, a truly perfect silicon crystal would be totally useless, being both a poor conductor and a poor insulator. Dopant impurities change all that by allowing us to vary the conductivity by a dozen or so orders of magnitude and to form pn junctions, the mainstay of device structures. Most other defects can also play a benevolent role, depending on the objectives of the device designer. For example, an efficient recombination center is undesirable in devices that require long carrier lifetimes, but quite desirable in devices where excess carriers need to be eliminated rapidly.

Another example of benevolent defects are the oxygen impurities which are known to be responsible for strengthening the wafer by reducing warping. Finally, a variety of extended defects purposely introduced at the backside of wafers are helpful in "gettering" impurities from the active region. With increased miniaturization, however, processing times are reduced so that gettering impurities to the backside of the wafer is becoming less practical. The controlled introduction of defects in the interior of the wafer for gettering purposes (intrinsic or internal gettering) is, therefore, becoming a desirable alternative.

One of the main objectives of device processing is to learn how to control and manipulate the two classes of defects described above, namely how to avoid, eliminate or passivate undesirable process-induced defects, and how to harness a number of defects for useful work such as impurity gettering. These objectives are often accomplished by engineering and materials know-how, by trial and error which is guided by experience, macroscopic physics (e.g., thermodynamics, solutions of transport or kinetic rate equations, etc.), and, sometimes, microscopic understanding. As the number of processing steps increases, however, and dimensions are getting ever so smaller, microscopic understanding is becoming more essential to help make informed judgements about processing choices.

There is, however, a third class of defects that are not discussed very often. They are, in my opinion, the unsung heroes, that will, eventually, become the ultimate tool of the device engineer. They are the defects that are actually responsible for the process itself. Let us take, for example, the very elementary process of diffusing dopant impurities into the wafer by thermal means (conventional furnace). As is well-known, impurity diffusion is mediated by intrinsic defects such as vacancies, self-interstitials, etc., which are created thermally. The rate of diffusion is proportional to the concentration of these intrinsic defects which is given by $\exp(-H_F/kT)$, where H_F is the formation enthalpy of the defect, k is the Boltzmann constant, and T is the temperature. Clearly, by choosing the temperature and time of the process, one can control the diffusion depth. In actual practice, elaborate computer programs have been developed which predict the dopant profiles that will result from specific processing conditions. These programs usually make a number of tacit assumptions about the microscopic processes involved and introduce adjustable variables to account for effects of oxidation, impurity concentration, etc. Clearly, detailed microscopic understanding of the intrinsic defects that mediate impurity diffusion, and the specific defect-impurity reactions as functions of external conditions (e.g., oxidation) should have a significant impact on diffusion modeling.

Let us, now, examine an alternative method of incorporating dopant impurities, namely ion implantation followed by short-time annealing using an arc lamp. Under these conditions, it has been found that impurities diffuse with substantially lower activation energies than in normal thermal diffusion.[4] One natural explanation is that ion implantation introduces defects which are then present to mediate impurity activation and redistribution. These defects need not be the same as the ones that mediate normal thermal diffusion. They are certainly induced by the process (ion implantation) and, in some kind of feedback mechanism, they are actually responsible for producing the desirable outcome of the process, namely the desirable dopant profile. At the same time, some of these defects probably aggregate in particular ways giving rise to the undesirable "process-induced" defects. Clearly, microscopic understanding of the primary process-controlling defects, the particular defect-impurity reactions that cause impurity activation, impurity redistribution, and the formation of undesirable "process-induced" defects, including the rates of these reactions, would be a valuable guide in determining which process conditions (energy of the ion beam, duration of heat pulse, maximum temperature, etc) would accomplish the desired dopant profiles. In addition, once the reactions are understood, one might try to select particular reactions by supplying energy in a very specific mode, e.g., using a particular laser wavelength, a second ion beam, etc. Thus the detailed understanding of the process-controlling defcts and their reactions might form the foundations of a new generation of modeling programs for process engineering. At the risk of being overly optimistic, this example illustrates that manipulation of defects and defect reactions at the microscopic level is the ultimate tool of the engineer.

The importance of defects in mediating and controlling processes is not limited to the example described above. Other examples abound. Take, for example, reactive-ion etching of Si. If fluorine were to be merely adsorbed on the surface, etching would not occur. Fluorine must penetrate the network and break bonds, i.e. create defects, which are then responsible for the ejection of SiF_n molecules (n=1, 2, etc.). Clearly, microscopic understanding of the process requires understanding of the bond-breaking energetics and kinetics as a function of Fermi-level position, the energy of incoming ions, etc. Similarly, oxidation or silicide growth involves interdiffusion, which is also mediated by defects. A well-known consequence of both oxidation and silicide growth is the "snow-plow effect":[5] impurities are redistributed in the silicon substrate with activation energies that are considerably smaller than normal thermal activation energies, indicating that defects induced by the process, probably the very defects that mediate the primary process, are involved.

In summary, we find that there are undesirable process-induced defects, "benevolent" defects that can be created on purpose to achieve certain objectives, and process-controlling defects which mediate the process and lead to both the desirable effect and the undesirable "process-induced" defects. Microscopic understanding of all these defects is valuable as a guide for making informed processing choices.

MICROSCOPIC UNDERSTANDING OF DEFECTS AND DEFECT REACTIONS

It is not possible in the short space allotted here to give a comprehensive account of the current microscopic understanding of defects and defect reactions. A number of excellent reviews are available. The author's views about the current understanding of point defects can be found in a recent article.[6] Here, we will discuss briefly only three issues as they relate to VLSI: Electronically stimulated defect reactions, the current state of understanding of some extended defects (dislocations, oxygen precipitates), and the microscopic mechanisms of defect-mediated atomic diffusion processes.

Electronically Stimulated Defect Reactions

Reactions occur when energy is supplied to overcome reaction barriers. Under normal thermal processing, energy is supplied by the vibrating lattice which stores energy in the form of phonons. Thermal energy, however, cannot easily be supplied selectively. Normally, the entire wafer must be heated which may induce undesirable reactions to occur. As a result, one must make sure that the heat treatment required for a particular processing step does not destroy the structures built by the previous steps. It is, therefore, desirable to develop processes that do not require high temperatures. One way to accomplish such a goal is to supply the energy in an alternative way. For example, energy can be injected in the lattice by a laser or an ion beam. Such techniques are used widely in silicon processing, but the mechanisms by which the energy is channeled to induce specific reactions have not been investigated extensively. On the other hand, it has been known that external sources of energy such as laser or electron beams create excess electron-hole pairs which store energy. This energy can then be channeled to enhance the rate of particular reactions. Many such examples are known, especially in compound semiconductors.[7] In the case of Si, one clean example is the migration of interstitial Al, which can be created by electron irradiation of Al-doped p-type Si.[8] Interstitial Al migrates thermally in the channels with an activation energy of 1.3 eV. If the sample is irradiated by a laser, however, the activation energy is reduced by almost 1 eV.[9] Theory has recently contributed to the microscopic understanding of the observed reduction in activation energy: In the presence of excess carriers, Al captures one or two electrons and temporarily changes its charge state. In the new charge state, the migration barrier is smaller by about 1 eV.[10] Such charge-state changes can actually result in athermal migration, as in the Bourgoin-Corbett mechanism[11] where a migration barrier is overcome completely by energy supplied from the successive capture of minority and majority carriers. The most celebrated example of athermal migration in Si is the case of the self-interstitial: Indirect but convincing evidence[8] has revealed that, under ionizing radiation, self-interstitials migrate athermally. Theoretical calculations have demonstrated that such migration is indeed possible along several paths.[12,13]

Recently, Küsters and Alexander found that dislocation velocities are enhanced by illumination (photoplastic effect).[14] The activation energy for dislocation motion at a particular value of stress was found to be reduced by about 0.7 eV. The enhancement has been attributed to the recombination of optically created excess carriers. The particular mechanism for converting the recombination energy to excess dislocation kinetic energy is not yet determined, however.

Electronic stimulation of defect reactions has not explicitly been exploited in VLSI Si processing, but the effect undoubtedly occurs whenever ionizing radiation is used. Detailed microscopic understanding may reveal such reactions and lead to novel ways of manipulating defects and reactions to accomplish desired objectives.

Dislocations

Dislocations have been studied for decades and many books and review articles have been written about their properties.[15,16] Detailed microscopic understanding is quite incomplete, however. Even though high-resolution electron microscopy (HREM) is now capable of 2-3 Å resolution, single columns of atoms cannot yet be resolved. As a result, the structure of the dislocation core is still not completely known. Indeed, new information obtained from more refined resolution often reveals new disparities between observations and theoretical models.[17] Theoretical methods are currently capable of calculating the energy levels of particular dislocation geometries,[18] but cannot yet determine the atomic positions by minimizing the total energy. Attempts at energy minimization have been made,[18] but the semiempirical models of force constants that are being used are not suited to describe the reconstruction of dangling bonds.

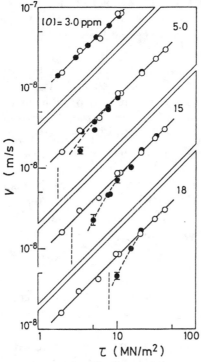

Fig. 1. The velocity of 60° dislocations as a function of stress at 647° C for different concentrations of oxygen. The open circles are for virtually oxygen-free float-zone Si. From Sumino, Ref. 19.

One of the more interesting properties of dislocations is their interactions with oxygen and other impurities. Such interactions have been extensively utilized in device production technology. Apparently, oxygen is capable of pinning dislocations, thus reducing the occurrence of slip and warping of Si wafers. Dislocations are also effective in gettering impurities, perhaps because of electrically active sites which attract charged impurities. The microscopic mechanisms underlying these effects are not, however, understood in detail. Measurements of dislocation velocities as a function of stress, temperature, and impurity concentration have been very valuable in providing information about both the dislocation core structure and dislocation-impurity interactions. For example, Fig. 1 shows results reported by Sumino[19] for the velocity of 60° partial dislocations at 647° C in silicon containing different concentrations of oxygen. The open circles are similar measurements in high-purity float-zone Si which may be considered oxygen-free. The figure shows that, as the oxygen content is increased, one must apply larger stresses in order to achieve the same velocity. Conversely, oxygen is capable of immobilizing dislocations for stresses up to a critical point (dashed vertical lines in Fig. 1), which depends on the oxygen content.

In contrast to oxygen, donor impurities tend to enhance dislocation mobility, whereas the effect of acceptor impurities seems to be in dispute.[19] As such studies continue and provide pieces of the puzzle, the microscopic picture is evolving with further help from theory. For example, kinks and kink pairs are usually considered responsible for dislocation motion, but the details of the mechanisms have not been completely determined.[16]

New information about the properties of dislocation has recently been provided by studies of the photoplastic effect, i.e., the enhancement of dislocation motion by illumination.[14] These observations probe directly the energy levels of dislocations.

Oxygen precipitates

Oxygen precipitates have attracted considerable attention because of the role they play in device-grade silicon. Several different stages that depend on the annealing temperature were recognized, but the nature of the precipitates remained unclear. Very recently, Bourret, Thibault-Desseaux, and Seidman[20] reported extensive HREM studies which appear to unravel many questions about the morphology of precipitates. They found that, in stage A (650° C) there exist long <011> ribbonlike defects which are SiO_2 microcrystals having the structure of coesite, a high-pressure SiO_2 polymorph. In addition, there are "black dots" which contain considerable amounts of oxygen, but whose structure could not be described in detail. In stage B (870° C), the black dots evolve into square platelets made up of amorphous silica. These platelets are quite thin (~15 Å). In stage C (1100° C), colonies of precipitates form stacking fault loops, which have been studied extensively by many. The results of Bourret et al.[20] give a great deal of insight into the nucleation and growth mechanisms of oxygen precipitates, but definite conclusions cannot yet be reached. One fascinating suggestion is that the so-called oxygen-related thermal donors (which appear after 350° C anneals)[21] may be the nucleation sites for larger oxygen precipitates. These ideas are currently been probed by more experiments and theory,[22] but definitive answers still remain elusive.

Defect-Mediated Atomic Diffusion:

The diffusion of Si atoms and dopant impurities (P, As, B, etc.) in Si is known to be mediated by intrinsic defects which are created thermally. The nature of these intrinsic defects has been the subject of intense controversy over the last 15 years.[23] There have been advocates of vacancies, divacancies, self-interstitials, and other more extended defects, but no consensus was achieved. There were two main difficulties. First, most of the arguments in support of particular mechanisms were based on interpretation of indirect observations for which other, often unrelated assumptions had to be made. Second, reliable theoretical calculations of some of the key quantities (e.g., formation and migration enthalpies of intrinsic defects) were not available.

Recently, Car, Kelly, Oshiyama, and Pantelides[13] reported parameter-free calculations of formation and migration energies of vacancies and self-interstitials in Si, providing, for the first time a firm theoretical framework for the interpretation of diffusion data. The new theoretical results contrast sharply with many of the earlier assumptions and estimates, but provide a consistent interpretation of the key data, reconciling apparent inconsistencies. In order to describe the main conclusions of that work, we first briefly describe the key experimental data and their prevailing interpretation.

The self-diffusion activation energy has been measured by many workers, but the resulting values vary between 4.1 and 5.1 eV.[23] If only one defect were mediating self-diffusion, the activation energy would be simply given by $Q = H_F + H_M$, where H_F and H_M are the formation and migration enthalpies of the defect. Independent information about migration energies was obtained in the 1960's by Watkins[8] who created vacancies and interstitials by electron irradiation and monitored their annealing kinetics. It was found that migration energies were quite small: the vacancy was isolated and found to migrate with an activation energy of ~0.3 eV; the interstitial could not be isolated, but it was concluded that, under the influence of electron irradiation, it migrated athermally. This information implied that formation energies of vacancies and interstitials would have to be large, of order 4 eV, in order to account for the observed high-temperature self-diffusion activation energy. Such a result conflicted with theoretical estimates of the vacancy formation energy which ranged about 2.5 eV and the belief that formation and migration energies ought to be comparable. In 1968, Seeger and Chik[24] proposed that the high-temperature defects are "extended", i.e., small amorphous bubbles, generally tetrahedrally coordinated, but having one more (interstitial) or one fewer (vacancy) atoms than would be present in the same region in the perfect crystal. The idea was suggested by the large values of the preexponential of the diffusion coefficient which is related to the entropy of formation of the defect. It was further argued that the "extended" interstitials can account for the low- and high-temperature data as follows: At low-temperatures, interstitials have a simple form and small migration barriers that can be overcome athermally by the Bourgoin-Corbett mechanism[9]. At high temperatures, they "spread out" and have larger migration energies. No detailed motional models were proposed, however.

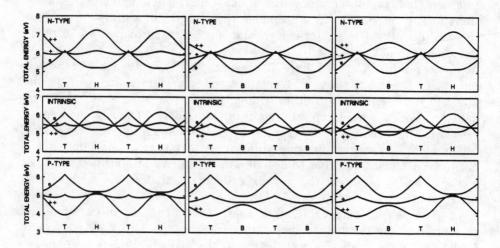

Fig. 2. Total energy of the self-interstitial in Si (from Refs. 13, 25) along three different paths, in three different charge states, for three different positions of the Fermi level. T and H are the tetrahedral and hexagonal sites in the low-density channels and B is the bond-centered configuration. The TBTH path correspons to motion along the (111) axis with the extra atom continuously exchanging sites with the next Si atom at a normal atomic site. Athermal migration is possible along all the paths shown by successively capturing one or two electrons and one or two holes. This figure illustrates the detail by which current theoretical calculations can trace the total energies of defects in Si.

The new theoretical results[13] provide a very different picture, which is simpler and consistent with observations without the need to invoke a change in the character of intrinsic defects as a function of temperature. It was found that both vacancies and self-interstitials have formation energies perfect crystal. The idea was suggested by the large values of the preexponential of the diffusion coefficient which is related to the entropy of formation of the defect. It was further argued that the "extended" interstitials can account for the low- and high-temperature data as follows: At low-temperatures, interstitials have a simple form and small migration barriers that can be overcome athermally by the Bourgoin-Corbett mechanism[9]. At high temperatures, they "spread out" and have larger migration energies. No detailed motional models were proposed, however.

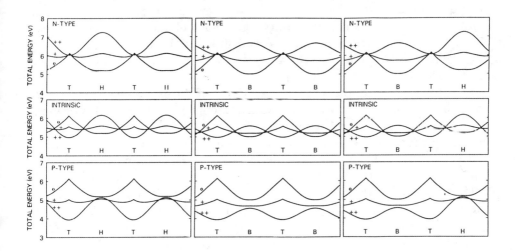

Total energy of the self-interstitial in Si (from Refs. 13, 25) along three different paths, in three different charge states, for three different positions of the Fermi level. T and H are the tetrahedral and hexagonal sites in the low-density channels and B is the bond-centered configuration. The TBTH path corresponds to motion along the (111) axis with the extra atom continuously exchanging sites with the next Si atom at a normal atomic site. Athermal migration is possible along all the paths shown by successively capturing one or two electrons and one or two holes. This figure illustrates the detail by which current theoretical calculations can trace the total energies of defects in Si.

The new theoretical results[13] provide a very different picture, which is simpler and consistent with observations without the need to invoke a change in the character of intrinsic defects as a function of temperature. It was found that both vacancies and self-interstitials have formation energies of order 5 eV.[25] The migration energies of vacancies could not be calculated, but the relevant migration energies of the self-interstitial were indeed found to be small (~0.5 eV) and possible to be overcome athermally (Fig. 2). Thus, low- and high-temperature data were reconciled in a simple and natural way. At low temperatures, vacancies and interstitials are created by electron irradiation and have small migration barriers, of order 0.5 eV. At high temperatures, self-diffusion is limited primarily by the need to create vacancies and self-interstitials thermally, which costs roughly 4.5-5 eV (theoretical error bar: of order 0.5 eV). The net self-diffusion activation energy is, therefore, of order 5 eV, as observed. Furthermore, the large entropy indicated by the observed large values of the preexponential can be accounted by the fact that the self-interstitial exists in several very different configurations and charge states with roughly the same formation energy.[12,13]

Since the new theoretical results on self-diffusion mechanisms were published, similar calculations relevant for dopant impurity diffusion have been carried out and are reported in Ref. 26. These results indicate that the activation energy of impurity diffusion is smaller than the activation energy of self-diffusion because the intrinsic defects (vacancies or interstitials) pair with the impurity. Detailed mechanisms for the migration of defect-impurity pairs have been described. In addition, another significant reaction occurs: Thermally created self-interstitials kick out substitutional impurities into the channels where they diffuse very efficiently, as found earlier for the self-interstitial.[12,13] Calculations of the activation energies of all these processes for P and Al yield numbers in overall agreement with experimental observations. These new results can now form the basis for resolving many of the existing conflicts regarding impurity diffusion,[23] and the fascinating results obtained from rapid thermal annealing of ion-implanted dopants.[4]

Acknowledgement: This work was supported in part by ONR Contract No. N00014-80-C-0679. The author would like to thank G. S. Oehrlein for valuable discussions.

REFERENCES

1. S. J. Fonash, this volume.
2. F. Shimura, this volume.
3. S. Lyon, this volume.

4. R. Kalish, T. O. Sedgwick, S. Mader, and S. Shatas, Appl. Phys. Lett. **44**, 107 (1984).
5. See, e.g., M. Wittmer and K. N. Tu, Phys. Rev. B **29**, 2010 (1984).
6. S. T. Pantelides, in *Deep Centers in Semiconductors*, edited by S. T. Pantelides, (Gordon and Breach, New York, in press).
7. See, e.g., L. C. Kimerling, Solid State Electron. **21**, 1391 (1978).
8. For a review of the early work, see G. D. Watkins, Inst. of Phys. Conf. Ser. No 23, 1 (1975).
9. J. R. Troxell, A. P. Chatterjee, G. D. Watkins, and L. C. Kimerling, Phys. Rev. B **19**, 5336 (1979).
10. G. A. Baraff, M. Schluter, and G. Allan, Phys. Rev. Lett. **50**, 739 (1983).
11. J. Bourgoin and J. W. Corbett, Phys. Lett. **38A**, 135 (1972).
12. Y. Bar-Yam and J. D. Joannopoulos, Phys. Rev. Lett. **52**, 1129 (1984); Phys. Rev. B **30**, 2216 (1984).
13. R. Car, P. J. Kelly, A. Oshiyama, and S. T. Pantelides, Phys. Rev. Let. **52**, 1814 (1984).
14. K. H. Küsters and H. Alexanders, Physica **116B** 594 (1983).
15. For a recent review, see H. J. Queisser, in *Defects in Semiconductors*, edited by S. Mahajan and J. W. Corbett, (Elsevier, New York, 1983).
16. A collection of timely articles may be found in J. de Physique Coll. C4, **44** (1983).
17. J. L. Hutchinson, in Ref. 15, p. 3.
18. S. Marklund, in Ref. 15, p. 25; J. R. Chelikowsky and J. C. H. Spence, Phys. Rev. B **30**, 694 (1984).
19. K. Sumino, in Ref. 15, p. 197.
20. A. Bourret, J. Thibault-Desseaux, and D. N. Seidman, J. Appl. Phys. **55**, 825 (1984).
21. G. S. Oehrlein, J. Appl. Phys. **54**, 5453 (1983), and references therein.
22. Several papers may be found in the Proceedings of the 13th International Conference on Defects in Semiconductors, to be published as a special issue of the J. Electron. Mater.
23. For recent reviews of atomic diffusion in Si, see W. Frank, Festkörperprobleme **21**, 221 (1982); W. Frank, U. Gösele, H. Mehrer, and A. Seeger, in *Diffusion in Solids II*, edited by A. S. Nowick and G. Murch, (Academic, New York, in press).
24. A. Seeger and K. P. Chik, Phys. Stat. Solid. **29**, 455 (1968).
25. Updated values for formation enthalpies, based on more extensive calculations are reported in Ref. 26. They are about 1 eV smaller than

than those reported in Ref. 13. The conclusions of Ref. 13 remain unchanged.
26. R. Car, P. J. Kelly, A. Oshiyama, and S. T. Pantelides, in Ref. 22, and to be published.

MICRODEFECT INTRODUCTION TO ENHANCE VLSI WAFER PROCESSING

Graydon B. Larrabee
Texas Instruments Incorporated, Dallas, Texas 75265

ABSTRACT

As minimum geometries approach sub-micron dimensions on VLSI devices and wafer sizes increase toward 200 mm, stringent requirements are placed on the wafer during device fabrication. Throughout all thermal processes the wafer flatness must be kept at one micron/cm local slope, dislocation related slip generation must be eliminated, impurities must be gettered both internally and externally and the electrical properties of the silicon in the device region must be enhanced. Wafer strengthening is accomplished through dislocation pinning using interstitial oxygen and small oxygen precipitates in Czochralski silicon. Gettering is accomplished using controlled microdefect formation. This is done internally using oxygen nucleation/precipitation microdefects in Czochralski silicon and externally with backside damage and polysilicon films. The size, density and morphology of the oxygen precipitate associated microdefects must be tailored to exactly match each VLSI device manufacturing process in order to simultaneously enhance both wafer strength and impurity gettering.

INTRODUCTION

Silicon devices for VLSI and for ULSI will place stringent requirements on the wafer during device processing. There is a clear trend for minimum geometry (design rules) shrinkage to continue and will appproach 0.5 micrometers in the 1990 timeframe. The current goals of VHSIC are 0.5 micrometers. These smaller device elements will result in storage of fewer electrons and will make DRAM memory cells more vulnerable to leakage and to upsets from alpha radiation. Physical defects such as decorated dislocations, precipitates, etc. will readily act as deep traps and cause significant device problems. CMOS devices will experience standby power problems that are aggravated by increased leakage currents. In bipolar devices, reductions in the sizes of p-n junctions will make them more vulnerable to leakage, degraded breakdown voltage and degraded transistor gain. Clearly the results of the presence of electrically active physical defects in the device region will have a devastating effect on the very small VLSI and ULSI devices of the 1990's.

Concomittant with these decreases in device geometries there will be an increase in wafer diameter from the current 125 to 150 mm up to 200 mm by 1990. Thermal processing of these larger diameter wafers will be a challenge and defect generation during thermal cycling will be difficult to circumvent. Processing of 150 mm wafers in today's processing environment indicates that slip generation is difficult to control, even at temperatures in the 950 to 1000°C range. A 200 mm wafer will have problems in the 850 to 950°C range. Thus it appears that defect generation will become a serious problem from the standpoint of the wafer and how the wafer is processed.

0094-243X/1220139-06 $3.00 Copyright 1984 American Institute of Physics

DISCUSSION AND REVIEW

Maximizing wafer strength can obviously be highly beneficial in controlling the generation of defects during thermal processing. Traditionally this has been accomplished in silicon integrated circuit manufacture through the use of Czochralski (CZ) silicon. CZ silicon, unlike float zone (FZ) silicon, which is not used in integrated circuit manufacture, contains interstitial oxygen that is far in excess of the solid solubility (Table I). This oxygen is introduced during crystal growth and comes from the oxygen supplied to the melt via dissolution of the quartz crucible. The oxygen strengthens the silicon wafer by inhibiting dislocation motion (pinning or locking of dislocations)[1,2]. The net result is the prevention of bow, warpage and slip formation during thermal processing[3,4].

Table I
Solid Solubility of Oxygen in Silicon at the Melting Point (MP) and at Various Wafer Fabrication Temperatures

Temperature °C	Solubility (old ASTM values) (Atoms/cm^3)	PPMA
1410 (MP)	2.6×10^{18}	52.0
1200	9.5×10^{17}	19.0
1100	5.2×10^{17}	10.4
1000	2.6×10^{17}	5.2
900	1.2×10^{17}	2.4

The exact oxygen species responsible for dislocation pinning is still under investigation. It is clear that oxygen nuclei (clusters of silicon-oxygen atoms) of less than a few hundred angstroms in size do not act as dislocation sources and appear to block dislocation motion[5,6]. Interstitial oxygen also has been shown to strengthen the wafer by precipitation on the dislocations thereby inhibiting motion[7,2]. In either case it is important to have interstitial oxygen and oxygen nuclei present in the wafer at all stages of device fabrication in order to minimize warpage and bow.

However, the role of oxygen is complex and wafer strength involves interstitial oxygen, oxygen nuclei and oxygen precipitates grown in the crystal lattice at high temperatures. During growth the precipitates can "punch out" prismatic loops of dislocations (Fig. 1) because of the large lattice misfit. Precipitates larger than a few thousand angstroms become dislocation sources. These dislocations are responsible for the decrease in wafer strength. There are direct correlations between the amount of oxygen precipitated and bow and warpage of a wafer[4,8].

These same precipitates appear to provide the internal gettering properties associated with CZ silicon. Therefore the amount of precipitation, and size distribution of precipitates, occuring during wafer fabrication must be closely controlled. There is an obvious tradeoff between maintaining wafer strength, vis-a-vis flatness, and precipitate growth for internal gettering. This is shown schematically in Fig. 2 where it can be seen that quantification is not attempted. From the earlier discussions it is probable that the optimum precipitate size is less than a few thousand angstrom but larger than a few hundred angstroms.

OXYGEN PRECIPITATION AND DISLOCATION GENERATION

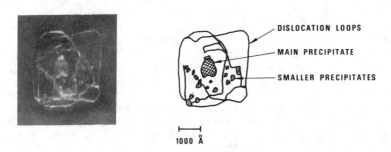

Fig. 1: TEM image and schematic of a platelet precipitate/defect complex.

EFFECT OF OXYGEN PRECIPITATE SIZE ON SLICE STRENGTH AND GETTERING EFFICIENCY

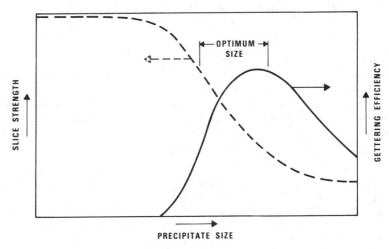

Fig. 2: Effect of oxygen precipitate size on wafer strength and gettering efficiency.

    The morphology of the oxygen precipitate and associated defects shown in Fig. 1 is typical of many platelet defects seen after the precipitation process. However, it is clear that there are other types of precipitate-defect clusters and these must also be comprehended in order to optimize internal gettering and wafer strength. There is a polyhedral type oxygen precipitate that is more typically < 600 Å in diameter but does not have the punched out dislocation network generally associated with the platelet defects. It is not obvious why one forms rather than the other, nor is it understood which is the "better" type for gettering or wafer strengthening. The initial nucleation process would be expected to play a dominant role in establishing the final morphology of the precipitate[9,10].

The nucleation process is either homogeneous or heterogeneous. There is not sufficient experimental data to conclusively point to one process or the other. In the case where the carbon levels are high, e.g., $>$ 5.0×10^{16}, the nucleation process is heterogeneous and the polyhedral morphology dominates the oxygen precipitates that form. However, when the carbon levels are low, e.g., $< 3.0 \times 10^{16}$ atoms/cc, the nucleation process appears to be homogeneous[11,12]. Some workers suggest that the nucleation process is still heterogeneous in the low carbon case, with the nucleation centers being other lattice defects. This latter position tends to be inconsistent with the experimental evidence that the nucleation centers can be eliminated using a high temperature treatment, e.g., 1300°C for 30 minutes. Following this high temperature treatment there is no oxygen precipitation after a typical 1000 to 1100°C precipitation thermal cycle. If the wafer is nucleated after the 1300°C treatment at 750°C, then there is precipitation with a typical precipitation thermal cycle. Significantly more research is required in the area of nucleation.

The device region obviously must be free of the precipitate/defect species. This is accomplished in bulk wafers by outdiffusing the oxygen at the wafer surface prior to precipitation to form a denuded zone[13,14]. When the denuding or outdiffusion is performed at temperatures in the 1100°C range, the interstitial oxygen concentration in the wafer can be as much as a factor of 10 above the solid solubility. Without nucleation, the thermodynamics favor outdiffusion and the surface concentration approaches the solid solubility of oxygen at that temperature. With this lower level of oxygen, precipitation cannot occur and the device region will be free of oxygen precipitate related defects. Denuded zones as deep as 50 micrometers are attainable. This same procedure is employed even if an epitaxial film is deposited on the bulk wafer. This ensures that stacking faults will not form in the epitaxial layer from oxygen precipitates that could form at the epitaxy wafer interface.

The gettering process associated with the precipitate/microdefect species is not well understood. Precipitation of metals on dislocations has been documented. There are large numbers of stacking faults associated with these microdefect species. Stacking faults form when there is an excess of silicon interstitials and a nucleation site (perhaps impurity precipitates or clusters). Both conditions are readily satisfied in CZ silicon wafer processing. There is increasing evidence that the silicon interstitial is playing a dominant role in the gettering process and there are many workers engaged in attempting to unravel the gettering process.

Backside damage can be an efficient gettering site. The mechanisms for backside gettering are probably the same as for internal gettering and are not well understood. Backside damage is implemented in many forms including sand blasting, scribing, ion implantation, laser damage, heavy phosphorous diffusion and polycrystalline silicon films. All rely upon the formation of stacking faults, misfit dislocations, etc. at the back surface of the wafer. The defect region extends up to a few micrometers into the back surface. However, one of the problems associated with extrinsic backside gettering is the propensity for the damage to be annealed out early in the wafer fabrication process. Thus it would only be effective for the first one or two thermal processing steps. The one exception to this scenerio is polysilicon backside gettering. In this case, apparently the stress induced by the polycrystalline film on the underlying wafer ensures that there are always misfit dislocations and stacking faults present. Extrinsic gettering is complementary to intrinsic oxygen precipitation gettering.

SUMMARY

The final product of the crystal growth and wafer fabrication process is a wafer that exactly matches the requirements of a specific VLSI or ULSI device and the manufacturing process used to make the device. Fig. 3 shows schematically a materials-system-silicon-wafer with all of the attributes necessary to optimize manufacturing yield and device performance. In this example oxygen concentration has been precisely introduced during crystal growth, e.g. $1.6 \times 10^{18}/cm^3$ $\pm$ 5%, and the carbon level kept less than 3×10^{16} atoms/cm^3. The wafer has been thermally processed in a manner that has formed a denuded zone, an epitaxial layer and internal nuclei of controlled size and number density. (The epitaxial layer would of course be omitted if the device was to be manufactured on a bulk wafer.) The wafer has controlled backside damage that matches the device manufacturing process. Low temperature thermal processing will become important in VLSI and ULSI device manufacture in order to minimize lateral diffusion of sub-micron structures. Controlled internal oxygen precipitation for internal gettering will be extremely difficult to achieve in low temperature wafer processing. Engineering microdefect introduction to enhance VLSI and ULSI wafer manufacture will be the key to success in the 1990's.

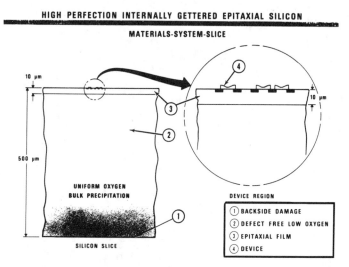

Fig. 3: Materials-system-wafer for VLSI and ULSI.

REFERENCES

1. K. Sumino and M. Imai, Phil. Mag. A, 47, 753 (1983).
2. H. Harada and K. Sumino, J. Appl. Phys., 53, 4838 (1982).
3. B. Leroy and C. Plougonven, J. Electrochem. Soc., 127, 961 (1980).
4. S. Takasu, et al., Jap. J. Appl. Phys., 20, 25 (1980).
5. K. Yasutake, et al., Appl. Phys. Lett., 37, 789 (1980).
6. L. Jastrzebski, IEEE Trans. Elect. Devices, ED-29, 475 (1982).
7. K. Sumino, H. Harada, and I. Yonenaga, Jap. J. of Appl. Phys., 19, L49 (1980).
8. S. Kishino, T. Aoshima, A. Yoshinaka, H. Shimizu, and M. Ono, Jap. J. Appl. Phys., 23, L9 (1984).
9. A. Bourret, J. Thibault-Desseaux, and D. N. Seidman, J. Appl. Phys., 55, 825 (1984).
10. G. S. Oehrlein and J. W. Corbett, Mat. Res. Soc. Proc., 14, 107 (1983).
11. H. L. Tsai, R. W. Carpenter, and J. D. Peng, In Press
12. R. Pinizzotto and S. Marks, Mat. Res. Soc. Proc., 14, 147 (1983).
13. D. Heck, R. E. Tressler, and J. Monkowski, J. Appl. Phys., 54, 5739 (1983).
14. I. Isomae, S. Aoki, and K. Watanabe, J. Appl. Phys., 55, 817 (1984).

DEFECT CONTROL AND UTILIZATION IN SILICON SUBSTRATES FOR VLSI

Xu Kang, Wang Weiling, Shi Zunlan,
Luo Guichang, He Dezhan, Shao Haiwen,
Shanghai Institute of Metallurgy, Academia Sinica
865 Chang Ning Road, Shanghai 200050, China

ABSTRACT

Some experimental results obtained from a study on defect control and utilization in silicon substrates for VLSI are reported and discussed.

INTRODUCTION

There is a variety of defects in silicon substrates, which can seriously impact IC performance and yield as well as reliability,[1-4] especially for VLSI circuits with decreased feature size. As the generation and mutual interaction of some defects such as unwanted metal contaminants and stacking faults (SF) during device fabrication are almost endless, defect control and utilization which can efficiently improve the material quality and device performance are of immense importance to VLSI circuits. This has led to the rapid development of a field of research referred to as "defect engineering" or "defect management" in recent years.[3-4]

In this study, we utilize the dynamic nature of defects to create a defect free region near the front surface, i.e. the denuded zone (DZ) for device structures, while isolating the potentional detrimental defects away from this region where they act as gettering sinks of unwanted defects introduced during device fabrication. The first methods we studied were chlorine oxidation gettering (COG)[5,6] and high temperature N_2 annealing (HTNA)[7] both using point defects. The second method was intrinsic gettering (IG)[8] using oxygen precipitation dislocation complexes (PDC) or related micro defects (MD). In the following some experimental results obtained are described and discussed.

EXPERIMENTAL PROCEDURE

The chlorine additive, TCE (C_2HCl_3), in COG was carried by 50 ml/min. N_2 in a O_2 flow of 1500 ml/min., while the N_2 flow during HTNA was 2000 ml/min; both were carried out at 1150°C.[11] In IG, either 2-step[9] or multi-step[10] annealing treatment is used.[12]

After pretreating, a number of diodes and MOS structures were fabricated. Then diodes leakage current I_L and MOS generation lifetime τ_g as well as defect density D were determined and compared with those of non-pretreated samples.

Unless otherwise mentioned, only boron doped p-type, (100) oriented, dislocation free CZ-Si wafers with 8-10 or 15-18Ω-cm and a diameter of 40 or 50 mm were used as the substrates. The initial oxygen and carbon contents of IG wafers were 0.6-1.0×10^{18}cm^{-3} and 1.0-10.0×10^{16}cm^{-3} respectively.

RESULTS AND DISCUSSION

1. COG and HTNA

Table I shows the MOS generation lifetimes τ_g, and the number of weak break-down points n_{wp} of COG ad HTNA treated wafers.

Table I $\tau_g (\mu s)$ and n_{wp} in TCE COG and HTNA treated wafers

	TCE COG			HTNA	
Sample No.	yes	no	Sample No.	yes	no
5	79.5	6.9	1	31.3	25.0
6	63.8	17.1	2	69.7	41.7
11	93.0	64.1	3	97.0	69.0
16	144.5	37.8	4	84.5	59.2
25	140.3	23.9	8	72.9	41.8
26	72.3	20.5	12	85.0	48.5
τAv. =	98.9 + 35.1	28.5 + 20.5	τAv. =	74.9 + 22.8	47.5 + 15.3
n_{wp} =	13.4 + 4.0	53.0 + 18.0	n_{wp} =	26.0 + 5.0	53.0 + 18.0

It is seen that the generation lifetime times τ_g for TCE COG and HTNA treated wafers are 3.5 and 1.5 times of those untreated samples, and the (number of) weak points is 1/4 and 1/2 of those untreated respectively. In addition, the diode yield of TCE COG wafers can be increased from 30.5% to 85.4%, approx. a factor of 3 higher than those of untreated.[11]

Apparently these improvements can be attributed to the gettering effects of Cl_2 with respect to the metal impurities and OSF nuclei, such as S-pits, scratches as well as A swirl and $[O_i]$.[13] Moreover, the chemical reaction of Cl_2 with Si at the growing Si/SiO_2 interface causes the generation of excess point defects, such as vacancies at the silicon surface, so that OSF are suppressed and even eliminated during the chlorine oxidation. Concurrently, due to out-diffusion of $[O_i]$, a defect free region, i.e. DZ can be formed. While in the bulk, also because of Cl-Si reaction, the vacancy content Si_V is higher than that of interstitial Si_I. This will be beneficial to the oxygen precipitation, according to Hu's proposal that oxygen precipitates are stimulated by Si_V.[14] Therefore, if the COG time is not too long, these bulk oxygen precipitates and their induced MD may function as gettering sinks, as in the case of IG discussed in the next section.

However, it should be noted that as this gettering effect depends strongly on temperature[15] and is poor for gettering of Au and Pt group metals,[16] it is better to combine it with other gettering methods, such as a heavily doped PSG layer or laser induced damage on the backside.[16]

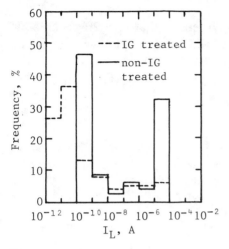 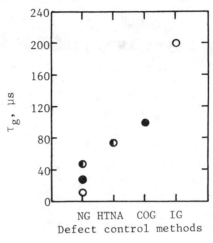

Fig. 1. Histogram of leakage current I_L from 292 IG and 185 non-IG treated diodes.

Fig. 2. MOS generation lifetime τ_g as a function of defect control methods.

Fig. 3(a-c) shows the surface denuded zone as well as bulk oxygen precipitates and related MD before and after CMOS 1K RAM fab-processing in an IG pretreated wafer. It is evident that the oxygen precipitates and their induced MD have grown bigger rather than shrunk during high temperature P-tub drive-in at about 1200°C for a period of 8-10 hours. Consequently the probe yield of CMOS 1K RAM can be upgraded by a factor of two. On the contrary, we can see from Fig. 3(d-f) that the oxygen precipitates and related MD are present throughout the surface active region and bulk after CMOS processing owing to non-IG pretreatment.

In comparison with the other gettering methods, the principal advantages of IG are the large capacity of the gettering sink and the shorter distance between the sink and active region of the devices as well as its ability to retain the gettering effect[2] for a long time. It should be noted, however, that too many oxygen precipitates in the IG wafer are harmful to the wafer flatness, in particular for the larger diameter wafer with fine feature sizes. For this reason, the oxygen content and its states in the starting wafers as well as subsequent thermal annealing sequences should be controlled appropriately.[2,3]

In summary, as the feature size scales down continuously with increasingly high integrity and low leakage current demands within a VLSI circuit, while each defect control or gettering method has its own capability, it seems that a single method is not always sufficient. Therefore a combination of two or even three defect control methods, such as IG complemented with mechanical or laser backside damage gettering[17] is essential in order to control all the defect problems

2. Intrinsic Gettering

The diode leakage current density J measured at $V_R = 20V$ is listed in Table II. Fig. 1 is the total I_L histogram. From these data it is shown that the IG wafers' J can be reduced to 10^{-2}-10^{-9} A/cm^2 with the best or peak value of $3\text{-}5\times10^{-10}$ A/cm^2, which are one to three orders of magnitude lower than those of non-IG treated samples. In addition, the defect density D_{20V} also shown in Table II can be decreased from 31.7 to 5.0 cm^{-2} on an average.

The measured MOS generation lifetimes τ_g and defect density $D_{0.8}$ are also listed in Table II. It is shown that after IG treating an average τ_g value of 205 μs can be obtained, which is one to two orders of magnitude larger than those of non-IG treated, and also higher than those of HTNA and COG wafers (see Fig. 2), while the defect density of thin gate oxides $D_{0.8}$ is decreased by a factor of 3 as compared with those untreated.

Table II Leakage current density J, generation lifetime
τ and defect density D in IG and non-IG treated
wafers with junction area of 1.44×10^{-2}cm^2

Sample No.	IG	J, A/cm^2		D, cm^{-2}*		τ_g, μs
		Total Av.	Peak Av.	D_{20V}	$D_{0.8}$	
3-1	no	2.4E(-4)	3.0E(-5)	64.2	67.0	2.5
3-8-5	yes	1.5E(-7)	2.3E(-8)	6.9	49.0	-
3-8-3	"	1.3E(-7)	1.5E(-7)	0	-	194.5
3-67-1	"	5.1E(-7)	2.5E(-8)	8.3	17.0	98.8
4-4	no	2.1E(-7)	2.5E(-8)	9.1	44.0	34.0
4-8-4	yes	2.6E(-9)	4.5E(-10)	3.6	3.5	137.0
4-5	no	4.2E(-4)	2.0E(-8)	26.4	45.9	4.0
4-8-5	yes	1.9E(-6)	4.6E(-9)	6.9	8.0	233.0
4-67-5	"	5.2E(-8)	3.0E(-9)	5.3	-	411.0
4-6	no	9.0E(-5)	2.2E(-7)	28.5	35.0	-
4-8-6	yes	4.1E(-8)	5.4E(-10)	6.3	19.0	-
4-67-6	"	2.6E(-9)	3.2E(-10)	6.9	-	-
4-13	no	5.2E(-5)	1.1E(-8)	30.5	48.6	-
4-8-13	yes	1.1E(-8)	1.5E(-9)	1.2	-	-
**5-1	no	-	-	-	-	6.0
5-67-x	yes	-	-	-	-	153.0

* D_{20V} - determined by $I_L > 1E(-5)$ A at $V_R=20V$ within a wafer;
  $D_{0.8}$ - determined by self-healing breakdown voltage of MOS
         structure $\leq 0.8V_{max}$. ;
** n-type (100) substrates, 5-8Ω-cm

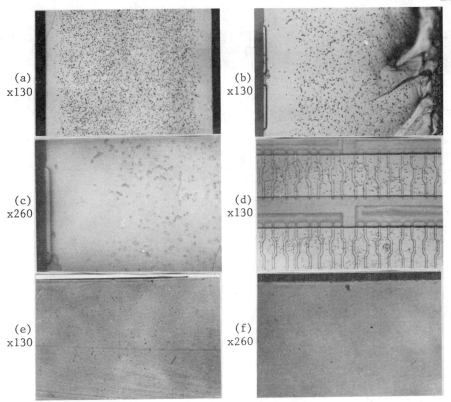

Fig. 3. Photos of IG (a-c) and non-IG (d-f) wafer. (a) surface DZ and bulk oxygen precipitates of IG wafers before and (b) after CMOS processing; (d) surface OSF and (e) bulk oxygen precipitates of a non-IG wafer after CMOS processing; (c) and (f) the same as (b) and (e) respectively, but at higher magnification.

occuring in silicon VLSI processes and achieve the desired superior performance in yield as well as reliability. Further studies on this subject are clearly necessary.

REFERENCES

1. D. Elwell, Progress Cryst. Grow. Charact. 4, No.4 (1981) 297.
2. L. Jastrzebesk, IEEE J. Solid State Cir., JSC-17, No.2 (1982) 105.
3. Hsu Kang, Semicond. Magz. (in Chinese), No.4 (1983) 12.
4. J.E. Lawrence, H.R. Huff, in VLSI Electronics, Vol.5 (N.G. Eingsruch, ed. Academic Press, N.Y. 1982) p.51.
5. R.S. Ronen, P.H. Robinson, J. Electrochem. Soc., 119 (1972) 747.
6. T. Hattori, ibid, 125 (1976) 945.
7. S. Kishino, Appl. Phys. Lett., 32 (1978) 1.
8. T.Y. Dan et al., Appl. Phys. Lett., 30 (1977) 175.

9. K. Nagasawa et al., Appl. Phys. Lett., $\underline{37}$ (1980) 622.
10. H. Tsuya et al., Jap. J. Apl. Phys., $\underline{20}$ (1981) L-31.
11. Hsu Kang et al., Keji Tongxin (in Chinese), No.2 (1983) 23.
12. Hsu Kang et al., CIE Conf. on Si-material and IC_s Tech. Digest (1983) (in Chinese), p.94.
13. C. Claeys et al., Semicond. Silicon (1981), Princeton, N.J., Electrochem. Soc., p.730.
14. S.M. Hu, J. Appl. Phys., $\underline{52}$ (1981) 3974.
15. C.W. Pearce et al., ibid (15), p.705.
16. L.E. Katz et al., J. Electrochem. Soc., $\underline{128}$ (1981) 620.
17. R.B. Swaroop, Solid State Technol., $\underline{26}$ (1983) 97.

Part VII. Devices

THE PHYSICS OF SCALED BIPOLAR DEVICES

T. H. Ning

IBM Thomas J. Watson Research Center, Yorktown Heights, NY 10598

ABSTRACT

The advent of the polysilicon emitter contact technology has solved the problem of insufficient current gain in scaled silicon bipolar transistors and allowed bipolar transistors to realize the performance advantage of miniaturization. The coordinated reduction of both the horizontal and vertical dimensions in bipolar scaling requires the base doping concentration to be increased appropriately to avoid punch-through. Recent progress in understanding the physics of these scaled bipolar transistors is discussed. Included in the discussion are the tunneling current in the heavily doped emitter-base junction, the minority-carrier transport in the thin-base region and in the heavily-doped emitter region, and the minority-charge storage in Schottky-barrier diodes.

INTRODUCTION

There have been major advances in silicon bipolar integrated circuit technology in recent years. Figure 1 is a schematic of a typical advanced silicon bipolar transistor [1], showing three key technology features, namely (i) the self-aligned polysilicon base contact, (ii) the deep trench isolation, and (iii) the polysilicon emitter contact. The polysilicon base contact and the trench isolation greatly reduce the device area and its parasitic capacitances, thus increasing the density and reducing the power-delay product of bipolar circuits [1-3]. The polysilicon emitter contact greatly improves the current gain [4] and allows bipolar transistors to be scaled down in dimension without suffering from the problems of insuficient current gain.

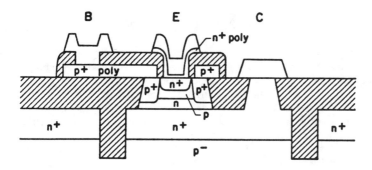

Fig. 1. Schematic of an advanced bipolar transistor (Ref. 1).

It has been shown [5,6] that in order to optimize the performance of bipolar ECL circuits in scaling, the base width should be reduced, while the emitter current density and the base doping concentration must be increased. The scaling rules as well as specific values for $0.25 \mu m$ optimized ECL circuits [6] are listed in Table 1.

The potential of such advanced bipolar devices for high-performance logic and memory applications [1,7] and for high-density VLSI applications [1-3] have been demonstrated. It has led to the growing interest in the physics of these advanced bipolar transistors, particularly in the scaling of these transistors and in the process-technology related device physics [4,8-17]. We will discuss the results of some of the studies on the physics of these scaled bipolar devices, and point out areas which either have not been studied or are not adequately understood.

Table 1.   Bipolar scaling rules and typical values for a 0.25 μm device optimized for ECL [6]. Here a = lithography dimension.

Lithography	a	0.25 μm
Base width W_B	$a^{0.8}$	25 nm
Base doping N_B	W_B^{-2}	7×10^{18} cm^{-3}
Current density J	a^{-2}	5×10^5 A/cm^2
Voltage	constant	

TUNNELING IN EMITTER-BASE JUNCTIONS

As the base doping is increased in scaling to avoid emitter-collector punch through [5,6], Zener tunneling current in the emitter-base junction in reverse bias increases rapidly. Not only is the peak doping concentration, but also the details of the emitter and base doping profiles are important in determining the tunneling current level. It was shown [11] that the tunneling current correlates very well with the maximum electric field in the junction. Since the maximum electric field can be calculated from the doping profiles, such correlation allows the tunneling I-V characteristics to be predicted.

For high-performance ECL circuits operating at approximately 1 mA and with small emitter-base reverse bias, tunneling current is not significant. However, for high-density circuits with more than 2-3 volt reverse bias, Zener tunneling current in the emitter-base junction must be examined carefully.

TRANSPORT IN THIN BASE REGION

As the base width is decreased, and the current density and the electric field are increased in constant-voltage scaling of bipolar devices [6], the usual assumption of equilibrium distribution is no longer valid for carrier transport in the base and collector regions. On the other hand, non-equilibrium transport, especially in regions with large doping concentration gradients, is extremely complicated to deal with rigorously. Recently, Cook [17] approached the problem by assuming the carrier temperature to be different from the lattice temperature and also an energy-dependent as well as field-dependent carrier mobility. As expected, the carrier drift velocity in this new model 'overshoots' that in the conventional model where equilibrium distribution is assumed. For the case of a very thin 30nm base, corresponding approximately to a 0.25μm bipolar device [6], the velocity overshoot is almost 50%, leading to a base transit time decrease of approximately 25%.

Cook's model thus does not predict appreciable circuit speed improvement from the velocity overshoot phenomenon in 0.5μm or larger devices. The 25% improvement in base transit time for a 0.25μm device is not negligible. The accuracy of the model, however, remains to be verified.

For the case of equilibrium distribution, the collector current density is given by [18]

$$J_c = \frac{qD_B(n_{ieff})^2}{W_B N_B} \exp(qV_{BE}/kT). \quad (1)$$

Here D_B is the minority-carrier diffusion coefficient in the base region, n_{ieff} is the effective intrinsic carrier concentration in the base region, W_B is the base width, and N_B is the average base doping concentration. It is a common practive to lump $D_B(n_{ieff})^2$ together and attribute any deviation from the easily measurable majority-carrier diffusion coefficient and the calculable $(n_i)^2$ to the band-gap narrowing effect [10]. In other words, band-gap narrowing is often used as a fitting parameter to account for the measured I-V characteristics.

In order to really understand the physics of the scaled bipolar transistors, it is important to separate D_B from the band-gap narrowing effect which contributes to the term $(n_{ieff})^2$. Recently, Dziewior and Silber [13] were successful in directly determining the minority-carrier diffusion coefficients using optical excitation techniques for doping ranges of 10^{17}-10^{19} cm^{-3}. This range should cover base doping concentrations of interest for advanced bipolar transistors but is not high enough for emitter doping concentrations.

HEAVY DOPING EFFECT IN BASE REGION

As the base doping concentration is increased to above 10^{18} cm^{-3} in scaling, the band-gap narrowing effect in the base region is no longer negligible [10]. While emitter band-gap narrowing increases the base current, and hence reduces the current gain, base band-gap narrowing increases the collector current, and hence increases the current gain. At low current densities, where base conductivity modulation is negligible, the heavy-doping effect in the base region of a submicron bipolar device can contribute 5-10X increase in current gain [10]. At high current densities where base conductivity modulation is significant, the electron and hole concentrations in the base and collector regions can be much larger than the impurity doping concentrations. The band-gap narrowing effect in such cases remains to be put on a rigorously physical footing.

MINORITY-CARRIER TRANSPORT IN EMITTER REGION

The transport of minority carriers in the heavily doped single-crystal emitter region has been the subject of extensive studies for many years, mostly in the context of solar cells [19]. As discussed earlier, the advanced scaled silicon bipolar transistors all have polysilicon emitter contacts. The transport of minority carriers across the polysilicon/silicon interface and in the heavily doped polysilicon emitter region is the subject of growing interest in recent years [4,15,16,20-22]. From the physics point of view, the problem is exceeding complicated because the properties of both the polysilicon layer and the polysilicon/silicon interface is a strong function of the details of the device fabrication process [15].

For devices where the polysilicon/silicon interfacial 'oxide' layer is thick, electron and hole tunneling through the oxide layer determine the transport properties [15,20]. For devices without significant interfacial oxide, the transport properties are determined by the transport properties in the heavily doped single-crystal and polysilicon emitter regions [4,15]. For most devices, it is likely that both interface and bulk transport must be invoked in order to explain the observed transport properties.

Many attempts [4,15,16,20-22] have been made, and it is likely many more attempts will be made, to construct comprehensive physical models of the polysilicon emitter contact. From the physics point of view, the greatest challenge remains the experimental determination of the minority-carrier lifetime and mobilities in the heavily doped silicon, particularly in the heavily doped polysilicon. The experiments by Dziewior and Schmid [23] give the Auger recombination lifetimes in heavily doped single-crystal silicon. There are no comparable data for heavily doped polysilicon. The experiments of Dziewior and Silber [13] give the minority-carrier mobilities in single-crystal silicon for doping concentrations up to about 10^{19} cm^{-3}. There are no comparable data for polysilicon. It remains a great challenge to physicists to determine the minority-carrier mobilities in polysilicon doped to greater than 10^{20} cm^{-3}.

SCHOTTKY-BARRIER DIODES

By using two-dimensional numerical simulation and advanced guard-ring SBD structures [24], Chuang and Wagner [12] were able to show that the minority-carrier quasi-Fermi level does not line up with the metal Fermi level at large forward bias. The implication is that the product

$$pn = n_i^2 \exp\left(q(\phi_p - \phi_n)/kT\right) \quad (2)$$

is much smaller than the prediction based on the equality between ϕ_p and the metal Fermi level. Consequently, the minority-charge storage in the SBD is much less than previously predicted.

SUMMARY AND DISCUSSION

The advanced scaled bipolar transistor technology offers ample opportunities and challenges to physicists who are looking for exciting physics in applied areas. The discussion in this paper represents but a very limited view of the important areas, concentrating primarily on fundamental transport properties. To those in the process technology community, the interests are often more in the process-related or process-dependent device physics. It is hoped that discussion like this will stimulate physicists to consider the challenges in the technology areas and pursuing a rewarding career solving technology-related physics problems.

REFERENCES

1. D.D. Tang, P.M. Solomon, T.H. Ning, R.D. Isaac, and R.E. Burger, IEEE J. Solid-State Circuits, SC-17, 925 (1982).
2. Y. Sugiyama, M. Suzuki, Y. Kobayashi, and T. Sudo, Proc. 1981 Custom Integrated Circuits Conf., 73 (1981).
3. K. Toyoda, M. Tanaka, H. Isogai, C. Ono, Y. Kawabe, and H. Goto, ISSCC Digest Tech. Papers, 108 (1983).
4. T.H. Ning and R.D. Isaac, IEEE Trans. Electron Devices, ED-27, 2051 (1980).
5. D.D. Tang and P.M. Solomon, IEEE J. Solid-State Circuits, SC-14, 679 (1979).
6. P.M. Solomon and D. D. Tang, ISSCC Digest Tech. Papers, 86 (1979).
7. F. Tokuyoshi, H. Takemura, T. Tashiro, S. Ohi, H. Shiraki, M. Nakamae, T. Kubota, and T. Nakamura, ISSCC Digest Tech. Papers, 220 (1984).
8. T.H. Ning, D.D. Tang, and P.M. Solomon, IEDM Tech. Digest, 61 (1980).
9. S. Gaur, IEEE Trans. Electron Devices, ED-26, 415 (1979).
10. D.D. Tang, IEEE Trans. Electron Devices, ED-27, 563 (1980)
11. J.M.C. Stork and R.D. Isaac, IEEE Trans. Electron Devices, ED-30, 1527 (1983).

12. C.T. Chuang and L.F. Wagner, presented at 1984 Device Research Conference, Santa Barbara, CA. Also submitted to IEEE Trans. Electron Devices.
13. J. Dziewior and D. Silber, Appl. Phys. Lett. 35, 170 (1979).
14. A. Neugroschel and F.A. Lindholm, Appl. Phys. Lett. 42, 176 (1983).
15. P. Ashburn and B. Soerwirdjo, IEEE Trans. Electron Devices, ED-31, 853 (1984).
16. Z. Yu, B. Ricco, and R.W. Dutton, IEEE Trans. Electron Devices, ED-31, 773 (1984).
17. R.W. Cook, IEEE Trans. Electron Devices, ED-30, 1103 (1983).
18. S.M. Sze, Physics of Semiconductor Devices, (Wiley-Interscience, NY 1969), p. 269.
19. See e.g. IEEE Trans. Electron Devices, Special Issue on Photovoltaics, ED-31, May (1984).
20. H.C. de Graaff and J.G. de Groot, IEEE Trans. Electron Devices, ED-26, 1771 (1979).
21. J.G. Fossum and M.A. Shibib, IEDM Tech. Digest, 280 (1980).
22. A.A. Eltoukhy and D.J. Roulston, IEEE Trans. Electron Devices, ED-29, 961 (1982).
23. J. Dziewior and W. Schmid, Appl. Phys. Lett. 31, 346 (1977).
24. C.T. Chuang, M. Arienzo, D.D. Tang, and R.D. Isaac, IEDM Tech. Digest. 666 (1983).

FUNDAMENTAL LIMITATIONS ON DRAM STORAGE CAPACITORS

W. P. Noble
IBM General Technology Division
Essex Junction, VT 05452, USA

W. W. Walker
Santa Clara, CA 95050, USA

ABSTRACT

If silicon DRAMs are to reach increasingly higher levels of integration, the area in which a single bit is stored will have to be continually reduced. In principle, this can be achieved in one or more of the following ways: reducing the thickness of the SiO_2 capacitor dielectric, introducing higher dielectric constant materials, or decreasing the amount of stored charge. This paper explores the physical limitations of these approaches as they apply to future 4-Mbit products. Analysis of previous design requirements indicates that the amount of charge stored will remain nearly unchanged while cell size must be reduced, resulting in insulator fields of ~ 8 MV/cm for planar capacitors. Insulator conduction and silicon impact ionization data, however, reveal that planar capacitors will not be viable, and three-dimensional capacitors such as trench or stacked structures must be introduced in order to meet acceptable leakage levels.

INTRODUCTION

In the course of MOS VLSI DRAM evolution, bit densities have increased from 16K to 1 Mbit per chip. The necessary decrease in storage capacitor size corresponding to this density increase has been achieved primarily by decreasing the thickness of the SiO_2 storage insulator, although capacitance enhancements have also been achieved by incorporating higher dielectric constant insulators and increasing the doping levels in the silicon. These approaches to increasing density will be pursued until physical limitations are encountered. This paper explores the known limiting mechanisms as they apply to the various process and design options, particularly in regard to their implications for a 4-Mbit product. This will be accomplished by first looking at the projected design requirements and then analyzing their compatibility with available data on insulator conduction, defects, insulator tunneling, and Si avalanche multiplication.

PRODUCT DESIGN REQUIREMENTS

The charge stored in a cell is determined by both the voltage which can be sensed by the sense amplifier and the transfer ratio of the cell and bit line, and an allowance for charge loss due to thermal leakage and ionizing radiation. The stored charge level for several generations of DRAM designs is shown in Table I. Although storage insulator thickness and power supply voltage change significantly, the amount of stored charge has remained at ∼ 150 fC, which suggests it will not change greatly in the future. Whereas ideal

Table I. Stored Charge Level

Bit Density	Voltage/Oxide Thickness	Charge
64 K	8.5V/45 nm	156 fC
256 K	5V/25 nm	180 fC
1 M	5V/15 nm	170 fC

scaling would lead us to decrease all dimensions as technology evolves, chip size increases (Fig. 1). Effective utilization of chip area requires keeping bit line length relatively constant while increasing the number of cells on the bit line. Thus, without a decrease in bit line capacitance, the stored charge level must stay up to maintain the transfer ratio.

Published data [1] on ionizing radiation soft error rate (SER) versus the critical level of stored charge, Q_{crit}, can be empirically fit to:

$$F_{rs} \propto \exp\left(- Q_{crit}/13 \text{ fC}\right), \qquad (1)$$

indicating that a 30 fC change in Q_{crit} increases SER by 10X.

As shown in Fig. 2, cell size has consistently decreased by 2.3X every 3 years, which points to a ∼ 13 μm^2 cell for a 4-Mbit chip. Assuming a constant ratio of node area to cell area, this implies a 5.2 μm^2 area in which to store 150 fC, or 29 fC/μm^2. The resultant field of 8 MV/cm in SiO_2 is much greater than before (1.6 - 3 MV/cm) and calls for a 4.4 μm SiO_2 storage insulator at the scaled voltage level of 3.5 V.

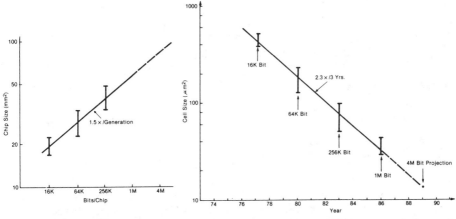

Figure 1.   Figure 2.

INSULATOR CONDUCTION LIMITS

The significance of the projected field strength can be judged by analyzing the existing data on insulator electrical properties. Since the relative dielectric constant (K) of SiO_2 is only 3.9, other insulators of higher dielectric constant have been explored. Si_3N_4 and Ta_2O_5 have dielectric constants of 7 and > 20 respectively. Although more charge can be stored at a given field strength with a higher dielectric constant material, conductivity generally increases. Data on the conductance of SiO_2 [4], Si_3N_4 [4], and Ta_2O_5 [5], indicates a Poole-Frenkel mechanism with the field (E) given by

$$J_{ins} = J_o \exp(B\sqrt{E}). \qquad (2)$$

Cell leakage due to insulator conduction can be ignored if it is an order of magnitude lower than that due to silicon leakage, or 0.3% charge loss per refresh interval. Applying Gauss' Law to this criterion establishes another relationship between insulator current and field:

$$J_{ins} \leq 0.76 \, K_{ins} \, \varepsilon_o \, E. \qquad (3)$$

The simultaneous solution to Equations (2) and (3) yield a limiting value of the field for each insulator. Table II shows the empirically determined values of the constants J_o and B, based on the data of references [4] and [5], and the calculated field limit for the three insulators. Also shown are the relative dielectric constants, K_{ins}, from which we can determine limiting stored charge

Table II. Insulator Conduction Limits

	J_o(A/cm)	B(cm/Mv)$^{1/2}$	E(Mv/cm)	K_{ins}	Q/A(fC/μm^2)
SiO_2	5.1×10^{-28}	17.7	7.9	3.9	27
Si_3N_4	9.2×10^{-18}	11.7	5.0	7.0	31
Ta_2O_5	4.1×10^{-15}	23.4	0.7	23.0	14

densities for these insulators. The charge density limits for SiO_2 and Si_3N_4 are nearly the same and approximately equal to the 29 fC/μm^2 projected product requirement. The conductance of Ta_2O_5 varies with preparation procedure, and further development may bring its charge density limit closer to SiO_2 and Si_3N_4. However, these results provide no basis to expect any fundamental advantage over SiO_2 for higher dielectric constant insulators.

INSULATOR TUNNELING

Below, a certain thickness quantum mechanical tunneling current through an insulator becomes significant. M. Fischetti [6] has used the WKB approximation to derive an expression for the insulator tunneling current:

$$J_{tun} = q\, v_t \frac{K_{ins}\, \epsilon_o\, V_{ins}^2}{K_{sil}\, T_{ins}\, kT} \exp\left\{ -C \frac{T_{ins}\, \phi_o^{3/2}}{V_{ins}} \left[1 - \left(1 - \frac{V_{ins}}{\phi_o}\right)^{3/2}\right]\right\} \quad (4)$$

where ϕ_o = barrier energy between the silicon and insulator conduction bands; V_{ins} = voltage across the insulator; T_{ins} = insulator thickness; v_t = electron thermal velocity; and C = physical constant containing electron effective mass and Planck's constant.

The field is not an independent variable in this tunneling equation. Using values of ϕ_o = 3.2 V and 1.9 V for SiO_2 and Si_3N_4, the thicknesses have been calculated which result in a current density of 2×10^{-6} A/cm^2, which is equivalent to that used to evaluate the insulator conduction limiting field. Calculations were done at fields corresponding to 17 and 10 fC/μm^2 (5 and 3 MV/cm in SiO_2). The results are shown in Table III. Thus, for a wide range of insulator fields, the calculated tunneling limited insulator thickness varies only slightly and shows only a weak dependence on insulator material. For SiO_2, this limiting thickness is only marginally less than the projected product requirement.

Table III. Tunneling Limited Thicknesses

	ϕ_o(V)	Q_{st}/A(fC/μm²)	Field(Mv/cm)	T_{ins}(nm)
SiO_2	3.2	29	8	3.8
SiO_2	3.2	17	5	3.0
SiO_2	3.2	10	3	2.8
Si_3N_4	1.9	29	4.5	4.5
Si_3N_4	1.9	17	2.8	4.1
Si_3N_4	1.9	10	1.7	3.7

SILICON DOPING CONSTRAINTS

An effective storage capacitor requires the smallest possible applied voltage across the insulator. This means that the depletion layer width within the silicon node in which Q_{st} is retained must be minimized. A reasonable requirement to meet is that the total storage capacitance of the silicon space charge capacitance in series with the insulator capacitance be diminished by no more than 10% of the insulator capacitance. This constrains the silicon surface depletion layer width and thus places a lower bound on doping given by

$$N \geq \frac{10}{T_{ins}} \frac{K_{ins}}{K_{sil}} \frac{Q_{st}}{qA} . \quad (5)$$

A second constraint is placed on the silicon node doping by the need to avoid impact ionization under the maximum field conditions, which would accelerate minority carriers toward the surface, creating electron-hole pairs in the surface depletion layer. The minority carriers at the surface will diffuse to the edges of the node and drift to the substrate, while majority carriers will drift back to the node bulk, thus constituting a leakage current. The maximum field exists at the surface and is determined by the stored charge density, independent of doping. For impact ionization to occur, the field must extend over a distance adequate to provide carriers with sufficient energy to undergo ionizing collisions. Since the rate of decrease of the field with distance from the surface is proportional to doping, the critical field E_{crit} required for impact ionization increases with doping density. Experimental results of Goetzberger and Nicollian [7] can be fit to

$$E_{crit} \sim N_B^{0.2} \, 2.3 \times 10^2. \tag{6}$$

Relating the field at the silicon surface to the stored charge density and expressing (6) as a doping limit, we get

$$N \geq 1.5 \times 10^{-12} \left(\frac{Q_{st}}{A K_{sil} \varepsilon_o} \right)^5. \tag{7}$$

Both Equations (5) and (7) place a lower bound on doping which increases with stored charge density. Assuming the insulator thicknesses of Table I, both relationships are plotted in Fig. 3. Up to

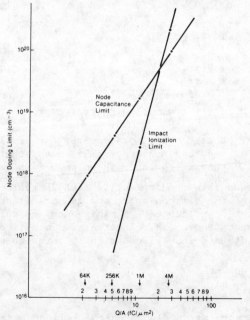

Figure 3.

1-Mbit density, node capacitance determines the lower bound on node doping and the necessary doping levels are readily achievable [8,9]. At the 4-Mbit design point, the limiting mechanism has become impact ionization and the required level of $2.5 \times 10^{20} \text{cm}^{-3}$ may well be unachievable.

DISCUSSION AND CONCLUSIONS

A 4-Mbit DRAM will require a cell size of $\sim 13 \ \mu m^2$ which, for a SiO_2 insulator, corresponds to a field of 8 MV/cm (i.e., 4.4 nm thickness for 3.5 V). Available data indicates that at the required field strengths, the insulator will be too conductive. Although insulators with higher dielectric constants may seem to offer an advantage, high field conductivity data show no actual benefit. Moreover, the projected insulator thickness (4.4 nm) is perilously close to the onset of tunneling breakdown.

The charge density requirement also encounters fundamental limitations. Although the node doping requirement imposed by practical space charge capacitance needs has always been a process constraint, the 4-Mbit density requires an even higher doping to avoid charge loss from impact ionization, and the required level of $\sim 2.5 \times 10^{20} \text{cm}^{-3}$ may well be unachievable.

It is thus evident that a change either in cell operation or fundamental capacitor design is needed for a 4-Mbit chip. Fields could be brought back into a safe range if the cell were operated so as to maintain the capacitor plate voltage at a level between the high and low node voltages. This necessitates circuitry to provide and control the plate voltage and would introduce an additional noise source. Extending the cell in the third dimension to produce more capacitance per planar area without increasing the field is an attractive alternative. Trench capacitors extending down into the substrate [2,8] and stacked capacitors on top of the active device areas [10] are already under consideration even for 1-Mbit designs, and we can expect these cell types to be prevalent in 4-Mbit chips.

REFERENCES

1. A. Mohsen et al, IEDM Tech. Dig., 616 (1982).
2. H. Sunami et al, IEEE Trans. Elect. Dev., ED-31, 6, 746 (June 1984).
3. R. Dennard et al, IEDM Tech. Dig., 168 (1972).
4. B. Deal et al, J. Electrochem. Soc., 3, 115 (1968).
5. G. Oehrlein, private communication.
6. M. Fischetti, private communication.
7. A. Goetzberger and E. Nicollian, Appl. Phys. Letters, 9, 12 (December 1966).
8. K. Minegishi et al, IEDM Tech. Dig., 319 (1983).
9. E. Adler et al, IEDM Tech. Dig., 327 (1983).
10. M. Taguchi et al, ISSCC Dig. of Tech. Papers, 100 (1984).

MONTE CARLO SIMULATION OF CONTACTS IN SUBMICRON DEVICES*

P. Lugli[+], U. Ravaioli, and D. K. Ferry
Center for Solid State Electronics Research
Arizona State University
Tempe, AZ 85287

ABSTRACT

We present an investigation of the transport through a metal-semiconductor Schottky barrier and its influence on the performance of a metal-n-n^+ submicron structure. This is achieved by using an Ensemble Monte Carlo (E.M.C.) simulation, in parallel with a numerical integration of Poisson's equation through the "collocation method". Hot-electron and space-charge effects are naturally included in this self-consistent particle-flow model. The tunneling mechanism from the semiconductor into the metal is also included in the simulation. It is found that the method provides the correct potential distribution in equilibrium and gives a detailed description of the physical processes involved in current transport through the interface when a bias is applied.

Metal-semiconductor contacts are of great importance in a number of semiconductor devices, whose applications range from high-speed logic to microwaves. As the dimensions of these devices reach the submicron limit, contacts become the limiting factor in the performance in the ballistic or quasi-ballistic mode of operation.[1] The standard theory of Schottky barriers[2] becomes less and less accurate when the devices have submicron dimensions, since the typical relaxation times in the system are comparable to the transit time through the device. The Ensemble Monte Carlo (E.M.C.) method[3] has been proven to be a very powerful tool in the study of hot electron effects and it has been widely applied to the analysis of submicron structures.[4]

The simulation of a 1-D metal-n-n^+ structure presents an interesting system in that the device is never charge neutral, except under flat-band condition. This is due to the presence of a depletion or an accumulation region near the interface. Since the value of the electric field at the two boundaries x = 0 and x = w (see Fig. 1) is related through Gauss' law to the net charge inside the device, it is necessary to allow the number of electrons simulated to vary during the simulation, as a constant number of electrons would give incorrect results. In this respect, the use of a cubic Hermite collocation method[5] to solve the 1-D Poisson's equation is crucial. In this method, the potential is represented by an interpolation consisting of a linear combination of cubic spline functions, whose coefficients are the values of potential and electric

---

*Supported by the Army Research Office and the Office of Naval Research
+Permanent address: Dipartimento di Fisica, Universitá di Modena, Italy

field at each mesh point. The choice of uneven mesh spacing is
allowed by the method without further complications. Boundary
conditions are imposed by setting the potential on the two
boundaries (at 0 and w), and interpolation constraints are set
by assigning the charge density in two collocation points (Gaussian
knots) inside each mesh interval. The values of potential and
electric field on the mesh points are simultaneously obtained by
directly solving a linear system of equations, whose size is twice
the number of mesh intervals. The collocation method has two main
advantages over the finite difference schemes: 1) for a given mesh
size h, the solutions obtained are $O(h^4)$ approximations to both
potential and field, compared to $O(h^2)$ for the potential in the
finite difference algorithm; 2) The value of the electric field
on the boundary is obtained without any loss of precision. The
knowledge of the electric field E_0 at the $x = 0$ boundary, gives
us directly the value of the current J through that interface,
through the relation $J = n^+e\mu E_0$, where n^+ is the impurity concentration in the highly doped region, e is the electron charge and μ
the low field mobility at that concentration. This relation
assumes that the electrons are in thermal equilibrium with the
lattice near the $x = 0$ boundary which is valid as long as the n^+
region width amounts to a few Debye lengths. The flux of electrons
out of the device through the $x = 0$ boundary is controlled by the
M.C. process. Thus, at every time step, the flux of electrons
in the opposite direction is given directly by the value of E
(i.e. of J) at the interface. This procedure allows us to update
constantly the numbers of electrons simulated, in accordance to
the field distribution inside the device. In this way, the interface at $x = 0$ is modelled as a perfectly injecting contact. The
metal contact at $x = w$ acts as an absorbing boundary: electrons
with energy sufficient to overcome the barrier are injected into
the metal (no image-force lowering is included in the simulation[6]).
The other conduction mechanism, from the semiconductor into the
metal, considered here, is provided by tunneling through the barrier.

The tunneling probability for an electron with energy ε at
a distance x from the interface is given by

$$T(\varepsilon) \cong \exp\{-2/\hbar \int_x^w dx \{2m^*[qV(x) - \varepsilon]\}^{\frac{1}{2}}\} \quad (1)$$

where m^* is the effective mass, V the potential seen by the electron, and $\hbar$ is Planck's constant. As a M.C. electron reaches the
barrier, the tunneling probability is calculated from (1) using a
parabolic least square interpolation of the potential $V(x)$ obtained from the solution of Poisson's equation. A random number
is then used to decide whether the electron will tunnel or not.
It is important to notice that no assumptions on the electron
distribution function near the contact or on the shape of the
potential barrier are needed. Quantum mechanical reflection
and field emission from the metal into the semiconductor could
be easily included along the same line.

In order to verify the validity of the self-consistent algorithm, we compare the equilibrium potential distribution

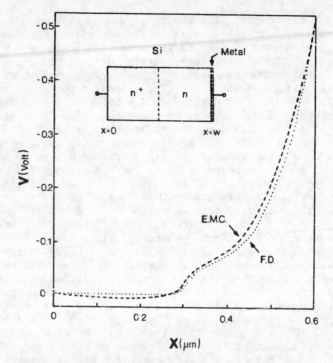

Fig. 1. Equilibrium potential distribution for the Schottky diode shown, as obtained from E.M.C. simulation (dashed line) and from Fermi-Dirac statistics (dotted line).

(reflecting the shape of the conduction band) obtained from the E.M.C. simulation with the result of an iterative process which combines the use of Fermi-Dirac statistics and Poisson's equation. The charge distribution across the device is obtained from the Fermi probability and then integrated to give the corresponding potential distribution. Convergence is achieved by relaxation of the intermediate solution for the charge through the imposition of a constant Fermi level throughout the device. The good agreement between the two methods (see Fig. 1) confirms the correctness of the E.M.C. (and of the boundary conditions used) simulating a Schottky diode. Fig. 1 corresponds to a 0.6μ - long Si/W structure at room temperature, with an impurity concentration of 10^{16} and $10^{17} cm^{-3}$ for the n and n^+ regions, respectively, and electric fields along the <111> distribution. The parameters used for the E.M.C. processes have been described elsewhere.[8]

Under high injection conditions, when a strong forward bias is applied, the barrier on the semiconductor side is greatly reduced and electrons are easily injected into the metal. The potential and the charge distributions corresponding to a band bending of $3k_BT$, where T is the lattice temperature, are shown in Fig. 2. It should be noticed that the carrier concentration in the vicinity of the barrier is much higher than the one obtained by Baccarani and

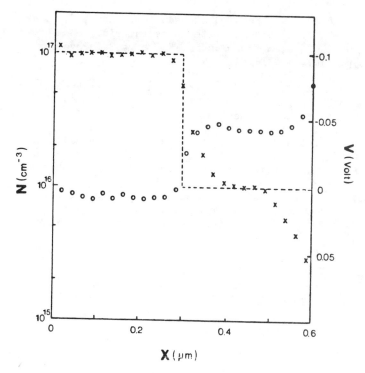

Fig. 2. Carrier (x) and potential (o) distribution under high injection conditions. The dashed line shows the doping profile.

Mazzone[9] assuming a parabolic fixed shape for the potential distribution. This is probably due to the tailing of the electron distribution from the highly doped region and to the consequent deviation of the potential distribution from the parabolic form characteristic of the depletion approximation.

At room temperature, for the electron densities considered here, current is provided solely by thermoionic processes, while tunneling is insignificant. Higher densities cause a narrowing of the barrier, thus increasing the tunneling probability. Our method indicates that significant tunnel currents start to appear for electron concentrations above $5 \times 10^{18} cm^{-3}$, in agreement with previous theoretical results.[2] This indicates that the method proposed here might be suitable also for the simulation of ohmic contacts.

The authors gratefully thank W. Porod and T. C. McGill for helpful discussion.

REFERENCES

1. D. K. Ferry, J. R. Barker and H. L. Grubin, IEEE Trans. Electron Dev., ED-28, 905, 1981.

2. E. H. Rhoderick, "Metal-Semiconductor Contacts", Clarendon Press, Oxford, 1980 and references therein.
3. P. Lugli, J. Zimmermann and D. K. Ferry, J. Physique, $\underline{42}$ (suppl. 10), 103, 1981.
4. C. Constant, in <u>Physics of Submicron Semiconductor Devices</u>, Ed. by H. L. Grubin, D. K. Ferry, and C. Jacoboni (Plenum, New York, in press).
5. P. M. Prenter, "Splines and variational methods", John Wiley & Son, New York, 1975.
6. A. Hartstein and Z. A. Weinberg, J. Phys. C: Solid State Phys., Vol. 11, p. L469, 1978.
7. J. L. Moll, "Physics of semiconductors", McGraw-Hill, New York, 1966.
8. P. Lugli, M.S. Thesis, Colorado State University, Fort Collins, CO, 1982; D. K. Ferry, Phys. Rev. $\underline{16}$, 1605, 1976.
9. G. Baccarani and A. M. Mazzone, Electron Lett., $\underline{12}$, 59, 1976.

THE INFLUENCE OF HOT CHANNEL ELECTRONS ON THE SURFACE POTENTIAL IN MOSFETs

D. Schmitt-Landsiedel, G. Dorda
Siemens AG, Central Research and Technology,
Otto-Hahn-Ring 6, D 8000 Munich 83, FRG

ABSTRACT

Hot carrier transport effects have been studied in Si-MOSFETs by means of capacitance-voltage measurements with applied drain voltages. The analysis of the experimental data shows that the main feature of hot electrons in the inversion layer is their redistribution in the conduction band and in correspondence with this the change of the voltage drop at the Si surface and the increase of the depletion charge.

INTRODUCTION

Upon decreasing the size of Si-MOSFETs, new phenomena become dominant in the function of the transistor. In particular, the transport process in the inversion layer is in most cases governed by hot electrons with their saturation velocity effect causing the breakdown of Ohm's law. Thus the investigation of the hot electrons in Si-MOSFETs becomes of great importance for small size devices.

In this paper we present capacitance-voltage (C-V) measurements which were performed on transistors applying drain voltage as a parameter. This procedure permits the investigation of hot carrier effects by comparing the capacitance data of small length MOSFETs (L<5 µm) with those of large lengths (L > 50 µm). The capacitance measurements were provided with a 1MHz Boonton bridge on MOSFETs with different channel lengths L and widths W. The oxide thickness was usually 33nm.

EXPERIMENTAL PROCEDURE AND ANALYSIS

Experimental results of normalized C-V data are shown on Fig. 1. Pronounced deviations from the so called classical behaviour of large scale transistors are evident. Our goal was to simulate the experimental curves by theoretical calculations. The theoretical model for the description of the C-V data with applied drain voltage is based on the theory of Pao and Sah[1]. The simulation of the hot electron curves was obtained by steps considering the saturation velocity and furthermore by incorporation of the electron temperature. As stated before[2], the

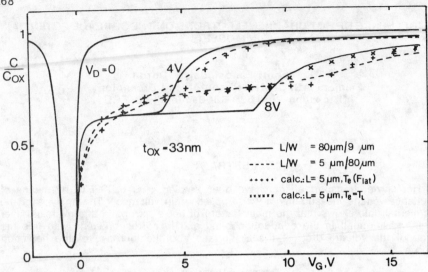

Fig. 1 Normalized gate capacitance C/C_{ox} vs. gate voltage V_G with the drain voltage V_D as a parameter for MOSFETs with long and short channel length.

capacitance increase in the saturation region ($V_G < V_D$) compared to the long-channel transistor, can be explained by the influence of the velocity saturation, whereas the introduction of the electronic temperature T_e with the equations

$$n = n_o \cdot \exp(-E/kT_e)$$
$$T_e = (1 + \tau_E \cdot q\,vF) \cdot T_1$$

of classical statistics gives a quite good description of the capacitance curve in the linear region ($V_G > V_D$). Here n is the electron concentration, E the energy, τ_E the energy relaxation time, v the velocity of the electrons, F the electric field and T_1 the lattice temperature.

It should be noted that even in the hot electron regime the motion of the electrons in the inversion layer is quantized. However, consideration of the quantum effects would result in significant calculation problems without any pronounced change in the overall description of the hot electron features. Thus all the evaluation of the redistribution of the electrons due to their increased energy, the calculation of their average distance x_{av} from the Si-SiO$_2$ interface as well as the changes in the potential drop at the Si surface have been provided by so called classical methods.

Increasing the electron temperature from $T_e = T_1$ to the value $T_e = 2T_1$ affects the distribution of the carrier concentration and the potential distribution at the Si surface as shown in Fig. 2, taking a constant carrier concentration $N_{inv} = 5 \times 10^{12}$ cm^{-2}. The two vertical markings in the concentration curves indicate the average distance x_{av} which indicates the increase of the electron distance from the Si surface due to their higher temperature. It should be emphasized that this effect appears despite the enlarged surface potential drop. This behaviour is a consequence of the raised energy of the electrons in the inversion layer which increases the probability of surmounting the potential difference to the substrate. Thus, a larger potential well must be built up to guarantee the constant carrier concentration N_{inv}. This can be realized only with an increased gate voltage.

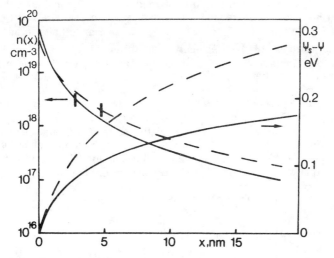

Fig.2 : The dependence of the electron concentration n and potential difference $\Psi_s - \Psi$ on the distance x from the Si-SiO$_2$ interface in thermal equilibrium (full lines) and with an increased electronic temperature $T_e = 2T_1$ (dotted lines) at constant $N_{inv} = 5 \times 10^{12}$ cm^{-2}. $V_G(T_e = T_1) = 3.25V$, $V_G(T_e = 2T_1) = 4.7V$, $T_1 = 300 K$.

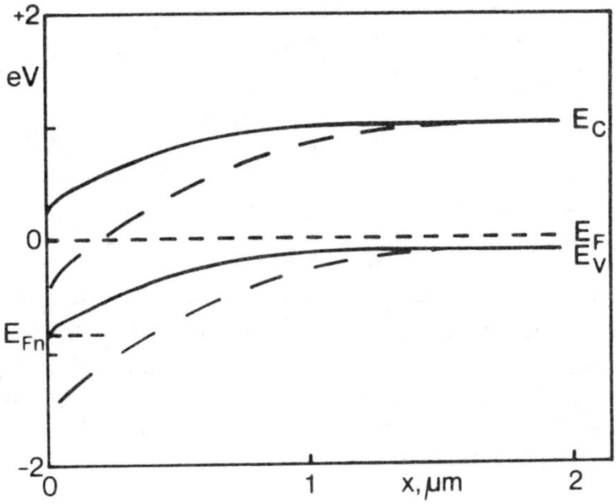

Fig. 3 : Band diagram at the Si surface in thermal equilibrium (full line) and with an increased electronic temperature $T_e = 2T_1$ (dotted line) at constant $N_{inv} = 5 \times 10^{12}$ cm^{-2}. Surface potential $\Psi_s(T_e = T_1) = 0.86V$, $\Psi_s(T_e = 2T_1) = 1.69V$, $T_1 = 300 K$.

Moreover, it is evident that with an enlarged surface potential drop the width of the depletion layer is also increased. This change of the surface band bending is illustrated in Fig. 3, again for $T_e = T_1$ and $T_e = 2T_1$. Simple considerations yield a factor for the increase of the depletion width given by $(T_e/T_1)^{1/2}$. This effect corresponds to an increased effective threshold voltage. An indication of this effect can be seen in Fig. 1 where the knee in the C-V curves at the saturation voltage is shifted to higher gate voltages.

For the correct description of hot electron behaviour in MOSFETs, in particular with respect to the C-V curves with applied V_D, the spatial variation of the electric field, of the mobile carrier concentration, of the surface potential distribution, etc. between source and drain must be determined. To give an idea of the changes in the electronic temperature T_e as well as in the average distance x_{av} for the hot electrons in a MOSFET an example is presented on Fig. 4. There is clearly

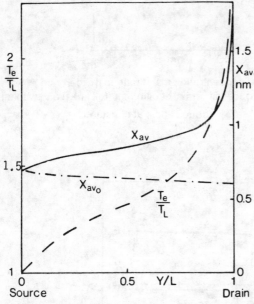

Fig 4 : Relative electronic temperature T_e/T_1 and average distances of electrons $x_{av}(T_e(F))$, x_{av0} ($T_e = T_1$) in dependence of location along the channel of the MOSFET. L/W = 0.55 μm/4 μm, t_{ox} = 10nm, VG = 5v, VD = 3v, T_1 = 300K.

seen the rapid increase of T_e near the drain region. On the basis of this T_e dependence the related average distance x_{av} is given, as well as for comparison the distance x_{av0} calculated without consideration of T_e, i.e. when taking $T_e = T_1$. This deviation illustrates very impressively how large changes in the distribution of the electrons in the hot electron regime can be possible, resulting also in modifications of the field distribution and current flow.

SUMMARY

In conclusion it should be noted that the appearance of hot carriers in the inversion layer of MOSFETs results not only in an increased average distance x_{av} of these carriers, but also, as a consequence of this, in an enlarged surface voltage drop, combined with an increased depletion layer width.

REFERENCES

1. H.C. Pao and C.T. Sah, Solid-State Electron. <u>9</u>, 927 (1966).
2. D. Schmitt-Landsiedel and G. Dorda, Electronics Letters <u>18</u>, 1041 (1982).

PHYSICS AT THE LIMITS OF VLSI SCALING

P.M. Solomon

IBM Thomas J Watson Research Center

Yorktown Heights N.Y.

Abstract

The integrated circuit technology is progressing at an unprecedented pace, with a doubling in number of devices per chip every two years. Dimensions are continually being reduced, and no limit is imminent in lithography. Lines narrower than 10nm have been produced. Among semiconductor devices, the MOSFET is predicted to be scalable down to about 100nm channel length - indeed, 150nm channel length MOSFETs have been demonstrated. The bipolar transistor should be scalable to dimensions of the same order. This paper will discuss an attempt to push the limit even further than those projections and in doing so will focus on device physics problems associated with very small devices. Silicon MOSFETs and gallium arsenide heterojunction FETs will be discussed.

The concept of scaling and the resulting limits of semiconductor technology has has been discussed extensively in the past, and we will not attempt to review this entire literature here. Dennard et al [1] proposed a scaling concept for the Si MOSFET, then practised at a lithography of 5μm which has been followed, with minor variations, down to 0.25μm [2]. Hoenheissen and Mead [3,4] studied the limits of field effect transistors (FETs) and Bipolar transistors, based on largely bulk semiconductor properties. Solomon [5,6] proposed a scaling of bipolar transistors based on scaling of the Poissons equation, and predicted limits for the bipolar transistor and FET, and Barker and Ferry [7,8] explored quantum transport for small FET devices. The field effect transistor (FET) and the bipolar transistor have been shown to be scalable down to dimensions of the order of 0.1μm. Classical scaling does not encounter any limits down to these dimensions. Many classical limits which had been identified in the past, such as power dissipation or breakdown voltage, were merely limits associated with an extant set of technological constraints, rather than being fundamental. Most of the more severe (I hesitate to call them fundamental) limits eventually encountered are quantum mechanical in nature, intimately tied in with the band structure of the semiconductor. Bulk properties of the semiconductor, such as electrical and thermal conductivities, breakdown voltages, and carrier concentration vs. energy become less meaningful as devices are scaled to very small dimensions. The following discussion will confine itself mainly to FETs, although many of the concepts (certainly the classical scaling) are applicable to bipolar transistors. Emphasis will be placed on the new FETs, based on III-V materials, partly because the silicon MOSFET has been extensively dealt with in the literature, and because the III-V materials exhibit strong quantum effects at larger dimensions. Extension of device operation to low temperatures (77K) will often be mentioned. The advantages of this mode of operation for scaled FETs and for super-fast large systems [9,5] have long been appreciated. Complementary FET technology CMOS is not dealt with explicitly, although the concepts discussed here apply equally to p and n

0094-243X/84/1220172-09 $3.00 Copyright 1984 American Institute of Physics

channel FETs. The concept of an intrinsic FET will be introduced here. This limiting FET, which may not be technologically attainable, has properties such as transconductance and capacitance directly relatable to those of the parent semiconductor.

Before progressing to the scaling of devices, we will consider some length and time parameters of importance for the device operation. The FET operates on electrostatic principles and to obtain good voltage gain and well controlled characteristics it is essential for the conducting channel to be in very close proximity to the control gate. The condition is:

$$t_{ins} < L\varepsilon_{ins}/\varepsilon_s \qquad (1)$$

where L is the source-drain spacing of the FET, t_{ins} the thickness of the gate insulator ε_s and ε_{ins} the permitivities of the semiconductor and gate insulator respectively. The 'field free' source and drain regions of the FET must be larger than the relevant screening length (see below). For the 'single electron' concept to apply, and to have enough quantum states in the device to be able to treat it classically, the de Broglie wavelength of the electrons has to be considerable smaller than L. For the device to switch off, i.e. to have isolation between source and drain and gate, the quantum damping length (inverse of electron's imaginary k vector) has to be less than L and t_{ins} respectively. Once these constraints have been satisfied, the FET should work. The precise device current will depend on the details of the electron transport. The various time scales affecting electron transport are discussed in detail by Barker and Ferry [7][8]. For our purposes, the questions concerning transport through the device can be separated from the questions concerning the operability of the device, which are addressed above. For instance a transition from a scattering dominated regime at large dimensions to perhaps pure ballistic transport at the smallest dimensions should not to first order determine the operability or non-operability of the device. The time scale of the device is the transit time of electrons from source to drain, ~0.1ps for the smallest devices. This approaches the time scales of the optical phonon frequency, and of collision and plasma frequencies in heavily doped semiconductors, such as would be used in the source and drain regions of the FET. We would therefore expect plasma resonances to significantly affect the current transients and more detailed work in this field is needed. Actual rise times of circuit waveforms are on a scale of ~ 10× larger than the transit time so that the exact details of the temporal behavior on the time of a transit time are of little concern to the logic circuit designer. The transit time is important in determining circuit speed since device currents - hence rates of rise of voltages on capacitors - are proportional to the transit time. Another figure of merit for a FET is the transconductance per unit width. The circuit time constant in a logic circuit is given by g_m/C where g_m is the transconductance and C is the node capacitance. The transconductance per unit width of the FET is therefore a measure of how fast a small FET can charge the device, parasitic and wiring capacitances of the circuit.

Dealing first with classical scaling, it can easily be shown that the set of Maxwell's equations can be scaled at a constant wave impedance Z_0, with distance and time being scaled by a common factor λ. One is still free to choose an additional scaling factor κ for potentials. Under these circumstances the current density increases as κ/λ^2, the charge density as κ/λ^2 and the electric field as κ/λ. A requirement of this mode of scaling is for the electrical conductivities to increase as $1/\lambda$. This mode of scaling applies to any linear network, and more specifically to the network of wires on a semiconductor chip. This implies that the chip dimension scale with λ and that the chip function stay constant. Actually chip dimensions are increasing and chip function is becoming ever more complex. Memory chips differ from logic chips since for memory the interconnectivity of the individual cell remains constant, and in 'random' logic it increases with chip complexity. To handle the extra connectivity in logic and to handle the long interconnects, a wiring hierarchy of multiple wiring levels would have to be adopted [5][10]. For the long term, logic approaches involving a constant connectivity per cell would be favored. The requirement for scaling of electrical conductivities cannot be met without changing materials or temperature. Fortunately FET circuits have not required the use of the highest conductivity materials available. Polysilicon, used in the present day MOSFET circuits (at about 3μm dimensions) has a conductivity of about 1000μΩ-cm, compared with about 100μΩ-cm for a typical refractory silicide, and 2.7μΩ-cm for Al. An additional factor of 10 can be achieved with Al at 77K, and 'ultimately' superconducting wires could be used. At small dimensions, <100nm, size effects become important in determining the wire conductivity. Coupled noise i.e. electromagnetic coupling between wires and between wires and devices is a problem even in today's circuits. The solution to this problem is inherent in the correct scaling of Maxwell's equations. Again the problem of the longer wires remains; however, they are equivalent to the former off-chip wires. All we are doing is putting more of the original system onto a single chip. We see that circuits can be scaled a long way before being limited intrinsically by the wires, provided failure mechanisms, such as electromigration, [11] can be averted.

Thermal diffusion equations in the bulk semiconductor can be similarly scaled (except that time t is scaled as λ^2 rather than λ). Local temperatures differences will increase if voltages are not scaled as least as fast as $\kappa = \sqrt{\lambda}$ (square root of voltage scaling). Actually the thermal environment is not very restrictive for FETs, and using parameters for silicon, it is estimated that a 0.25μm channel length MOSFET, at a power supply voltage of 1V dissipating 60μW per μm of channel width [2], will experience a temperature rise, relative to the substrate, of about 0.5°C. For low temperature operation the thermal problem is even easier, since the thermal conductivity of silicon increases by 20× going to 77K. It appears that classical concepts of heat removal will not yield a limit to the FET, especially if $\kappa \propto \sqrt{\lambda}$ scaling is adopted. Removal of heat from the semiconductor chip to its environment is a problem even today, but not one of a fundamental nature.

Focussing on the FET itself, it has been shown [2] [12] that the Poisson equation for semiconductors can be scaled in a manner similar to the treatment of the Maxwells equations above. The proviso is that

$$(E_F - E_C)/kT \dot\propto \ln(\kappa/\lambda^2) \qquad (2)$$

in order to maintain the required carrier concentrations, and that the temperature should be reduced as κ. Here k is the Boltzmann constant, T the absolute temperature, E_F the quasi-Fermi energy for electrons and E_C the conduction band energy. $E_F < E_C$ for a non-degenerate semiconductor. For FETs operating at voltages $>> kT/e$ (and for non-degenerate carrier statistics, as explained below) neither of the above provisos is important. Temperatures would have to be reduced if power supply voltages <1V become necessary. Coupling the scaling constants to the drift-diffusion equation results in $t \propto \lambda^2/\kappa$. This result is similar to that derived for the thermal transport equations, but gives a stronger reduction in delays (the familiar $t \propto \lambda^2$ for constant voltage) than for the interconnect network. This means that the interconnects, off as well as on chip, would eventually limit delays to $t \propto \lambda$, if off-chip scaling kept pace with on-chip scaling. Actually the non-linear nature of the drift-diffusion equation will eventually manifest itself, as it does already in GaAs devices, limiting both drift and diffusion velocities to constant 'saturation' values; then $t \propto \lambda$, in tune with the interconnects. Note that the practical effect of this scaling is to require the reduction of oxide thickness in proportion to channel length. The limit on the thickness of the gate insulator is due to quantum-mechanical tunneling and is about 3nm for SiO_2. This will reflect back to the device design to determine the minimum channel length.

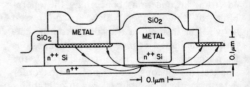

Fig. 1 A MOSFET design scaled near the limit, from Solomon [5]. The channel length is 100nm and the SiO_2 thickness 4nm.

A design for a MOSFET scaled close to the limit with a gate length of 0.1 μm (from ref. 5) is shown in Fig. 1.

The classical scaling rules above will begin to break down when the semiconductor becomes degenerate. While Boltzmann statistics apply, the above scaling procedure, with $\kappa \propto T$, implies that the Debye screening length in the semiconductor will scale with λ. When the semiconductor becomes degenerate the Thomas-Fermi screening length λ_{TF} applies:

$$\lambda_{TF} = \sqrt{\varepsilon h^2/4m^*e^2k_F} \qquad (3)$$

where m^* is the effective mass of the electron, h is Planck's constant, and k_F is the Fermi wave vector given by

$$k_F = (3\pi^2 n)^{1/3} \qquad (4)$$

and n is the electron concentration. This formula holds strictly for a single valley semiconductor like GaAs. Note that λ_{TF} is only a weak function of n. For GaAs λ_{TF} = 7.5nm for $n = 1 \times 10^{19}$ cm^{-3} and for silicon, which has a larger effective mass, λ_{TF} will be much smaller, about 1nm. The physical situation at heavy doping and high electron concentrations will be quite complicated [13] with the need to include various 'bandgap narrowing' effects. Non parabolicity effects, especially in the direct bandgap III-V semiconductors, will be pronounced, leading in general to a decrease of λ_{TF}. The magnitude of λ_{TF} will impose limits on the size of the 'neutral' regions of the device such as source and drain regions, and on inter-device isolation.

An assumption of the above scaling approach, which will be violated as dimensions are reduced, is that the transport properties are a function of local field and concentration gradients only. Non-local, non-equilibrium transport [14] is predicted, especially in III-V compound devices, as dimensions are reduced below 1μm. Impact ionization and hot-electron injection [15] effects are earliest to be affected, since they depend on the tails of an electron distribution which have to be built up across a considerable distance and potential span. Impact ionization, which was once identified as a limit to scaling [3], is not expected to be a problem in small devices since voltages will be comparable to or below the band-gap.

While silicon is the semiconductor almost exclusively used in integrated circuits, III-V compound semiconductors such as GaAs, (Ga,In)As, InP etc. are receiving increasing attention due to their superior transport properties. These are direct gap semiconductors and owe their superior transport properties to the small effective mass for the electrons in the conduction band as well as to the lack of intervalley scattering at low energies. This is in contrast to silicon, where strong scattering occurs between the multiple valleys. Band diagrams for Si and GaAs are shown in Figs 2 and 3.

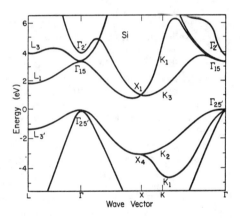

Fig. 2. Band diagram for silicon after Cohen and Bergstresser [33].

From the semiconductor band diagram one can derive a maximum group velocity for electrons as $dE/\hbar dk$ For Si one gets 6×10^7 cm/s and for GaAs 1.1×10^8 cm/s. Drift velocities in bulk semiconductors are determined from a solution of the Boltzmann equation [16] and given to first order by the drift diffusion equations. Major scattering mechanisms are impurity scattering and acoustic and optical phonon scattering, and inter-valley scattering. In GaAs, when conduction is in the central valley, acoustic phonon scattering is suppressed, resulting in much higher electron mobilities for GaAs compared to Si.

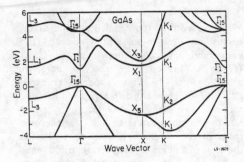

Fig. 3. Band diagram for GaAs after Cohen and Bergstresser [33].

At high electric fields carriers transfer to the satellite valleys in GaAs and velocities are reduced (as in Gunn devices). In doped channel FETs such as MESFETs impurity concentrations are increased rapidly as dimensions are reduced. The total number of impurities in the device are reduced (see below). When comparing a MESFET with a MODFET, which has an undoped channel, one has to compare the relative contributions of impurity to optical phonon scattering as dimensions are reduced. The scattering problem is further complicated in the case of transport in a FET by the 2-dimensional nature of the electron gas [17]. Experimentally, a saturated drift velocity at 300K of 1×10^7 cm/s is obtained in Si, and a peak velocity of about 2×10^7 cm/s in GaAs. A velocity as high as 3.8×10^7 cm/s has recently been observed [18] in GaAs modulation doped structures at 4.2K. Returning to the band diagrams, and the applicability of the Boltzmann equation, the de Broglie wavelength λ_B is given by

$$\lambda_B = 2\pi/\Delta k \qquad (5)$$

where Δk is the range of electron wave numbers in k space. This can be directly related to the lattice constant a, since the maximum range of k in the Brillouin zone is $\pm 2\pi/a$. As can be seen in the band diagram, the electrons can occupy a relatively large fraction of the Brillouin zone in Si, because of the broad minima (large effective mass) of the valleys, and de Broglie wavelengths may be only a few lattice constants. For GaAs, when electron energies are limited to below those of the satellite valley, fractional occupancy of the Brillouin zone is much smaller resulting in a larger de Broglie wavelength, about 6nm.

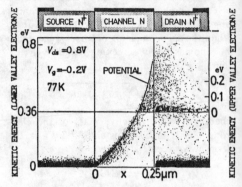

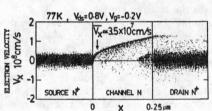

Fig. 4. Electron distributions in a $0.25\mu m$ channel length GaAs MESFET, obtained using Monte-Carlo simulations. From Awano et al [22]. (top) A cross-section of the MESFET. (center) The electron energy. (bottom) The longitudinal component of the electron velocity.

The drift diffusion equation is marginally valid for GaAs even with todays dimensions ($\sim 1\mu m$) although its use in silicon can be extended to much smaller dimensions. More accurate solutions of the Bolzmann equation by Monte-Carlo techniques [19,20] or by calculating the first and second order moments [21,22] have been used to simulate devices. These approaches are very computer intensive and still use many simplifying assumptions. Electron distributions in a GaAs MESFET as simulated by Awano et al [23] is shown in Fig. 4. This figure illustrates the ballistic regime, where the electron distribution is transported through the device with a minimal loss of momentum due to scattering. These simulations neglected electro-electron interactions which should be important at these electron concentrations. Awano's work shows clearly the effect of backscattering of electrons from the heavily doped drain region. This will tend to increase the charge storage in the device and degrade performance. Understanding of such effects will be ever more important as devices get smaller.

A problem in scaling investigated by Keyes [24] is potential fluctuations due to fluctuations in doping concentration. Potential (or field) changes, for instance, for threshold voltage control, are generated within semiconductors in p-n junctions, Schottky barriers, or heterojunctions by fields produced by ionized impurities in depletion layers. For uniform doping the potential change will be $enw^2/2\varepsilon$ where n is the doping concentration, w the depletion layer thickness, and ε the permittivity of the semiconductor. Scaling the potential and width by κ and λ respectively increases the doping by κ/λ^2. The number of impurities per cube of depletion layer width will decrease as $\kappa\lambda$ so that the relative potential fluctuation due to their random distribution will scale as $1/\sqrt{\kappa\lambda}$. For example, the $1/4\mu m$ device of Baccarani et al would have about 80 impurities per cube of depletion layer width. A solution to this problem, which might be needed at yet smaller dimensions, is suggested in ref. 5. Instead of the depletion layer width being defined by the doping, the width could be defined by an undoped region flanked by heavily doped regions or Schottky barriers. Potential steps could be defined by external voltages, or 'built-in' potentials could be created through the use of suitable work-function materials [25] and heterojunctions. For very small devices the 'number of electrons' in a FET channel may be very small. This problem is not the same as the problem of impurity fluctuations since the electrons are mobile. The number of filled states being small implies that λ_B (longitudinal) becomes comparable with the device length. In the limit the FET would tend to look like a coupled quantum well structure, the wells being the source and drain regions, coupled via a potential barrier which is varied by varying the gate field.

Certainly this is a topic warranting further theoretical work. Comparable FET type devices, on the basis of their operating principles, are the Si MOSFET and the GaAs MODFET (a review of the MODFET device is given in [26].) A band diagram for a MODFET is shown in Fig. 5. Like the MOSFET, electrons are confined in the channel in quantum sub-bands. The transconductance of these devices is inversely proportional to the thickness of the gate insulator (SiO_2 or (Al,Ga)As respectively).

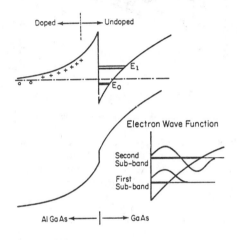

Fig. 5. Energy Band diagram of a MODFET device. From Solomon and Morkoc [26].

The tunneling limit for these insulators is about 4nm and 20nm respectively. Beyond the limits of specific insulators, one can describe a limiting FET having a zero thickness insulator, whose properties depend on the field effect properties of the semiconductor. In a FET an electric field is developed across the gate insulator by the gate voltage which, through Gauss law, modulates the channel surface charge. As the surface charge increases the Fermi level, at the semiconductor surface, rises with respect to the bands. By Kirchoff's law, the change in gate voltage is given by the sum of the changes of voltage across the insulator and the surface potential. In todays FETs the gate voltage changes by $\sim 1V$ for $\sim 0.1V$ change in surface potential, so that the surface potential change can be neglected in calculating the transconductance. In the limiting FET, the *entire* gate voltage is applied to the semiconductor. The information needed by a designer of this device is the relation between surface Fermi energy E_F and surface charge concentration n which is a consequence of the quantum sub-band structure.

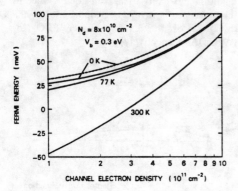

Fig. 6. Fermi level relative to the extrapolated GaAs conduction band edge, at the GaAs-(Al,Ga)As interface. The dashed curve gives the triangular barrier approximation. After Stern and Das Sarma [28].

Much work has been done in this area [27]. Recent data for a GaAs channel is shown in Fig. 6. from Stern and Das Sarma [28]. A good 'engineering' approximation to this relation is

$$E_F = k_s (e^2 hn/\sqrt{m^*}\varepsilon)^{2/3} \qquad (6)$$

where k_s is a constant which depends on the material and the well configuration. This formula resembles the simple variational formula [29] for the minimum energy of the first sub-band, but actually is a rough fit to the more involved derivation of the low temperature *Fermi Level* including multiple sub-bands and exchange and correlation effects. For calculations on Si [27] and GaAs [28] data one obtains $k_s = 0.47$ and 0.44 respectively. A surface capacitance C_s can be defined as

$$C_s = e^2 dn/dE_F \qquad (7)$$

and an inversion layer width w_{inv} by

$$w_{inv} = \varepsilon/C_s. \qquad (8)$$

The minimum device dimensions, by scaling, would be a multiple w_{inv}. The intrinsic transconductance per unit width is given by

$$g_m = C_s v \qquad (9)$$

and v is the average velocity of the carriers in the channel. To estimate v, we assume ballistic transport from source to drain. This assumption is probably justified for small GaAs devices (cf. Fig. 4.) but probably over-estimates the velocities for Si, so that maximum velocity of 1.3×10^7 cm/s was assumed [7]. The electron energy is made equal to the change in surface potential, as it would be for a logic circuit with equal gate and drain voltage swings. Maximum velocities are assumed to be limited to the velocity, in the central valley (assumed parabolic), at the energy of the minimum of the satellite valley, for the III-V semiconductors. As surface charge concentration increases the difference between the satellite valley minimum and the Fermi level is reduced. Results for E_F, v, w_{inv} and g_m are given in Tables I-IV for Si, GaAs and $In_{0.53}Ga_{0.47}As$. The satellite to central valley energy difference is larger in $In_{0.53}Ga_{0.47}As$ than in GaAs, 0.73eV vs. 0.36eV, and the effective mass ratio is smaller, 0.04 vs. 0.067.

Table I
Fermi Energy in meV vs.
surface concentration × 10^{12} cm^{-2}

n	Si	GaAs	$In_{0.53}Ga_{0.47}As$
0.5	30	60	70
1	40	100	110
2	74	150	180
5	140	280	340

Table II
Inversion layer thickness w_{inv} in nm

n	Si	GaAs	$In_{0.53}Ga_{0.47}As$
0.5	2.5	6.1	7.1
1	1.9	4.7	5.6
2	1.6	3.7	4.5
5	1.3	2.8	3.3

Table III
Average drift velocity $\times 10^7$ cm/s

n	Si	GaAs	$In_{0.53}Ga_{0.47}As$
0.5	1.1	2.9	4.0
1	1.3	3.6	5.1
2	1.3	2.6*	6.4
5	1.3	0.9*	8.0*

Table IV
Transconductance in S/cm

n	Si	GaAs	$In_{0.53}Ga_{0.47}As$
0.5	46	55	65
1	73	87	104
2	84	80*	165
5	110	37*	150*

* These values of velocity or transconductance are limited by the maximum kinetic energy of the electrons in the central valley at the energy corresponding to the bottom of the satellite valley.

The intrinsic transconductance values are far in excess of the best values in today's devices, by more than an order of magnitude. Although these numbers indicate that Si will be on a par with GaAs, in terms of transconductance; this would involve unreasonable thicknesses of SiO_2 in the Si case. An estimate of the insulator thickness needed for the FET to approach the 'intrinsic' FET is:

$$t_{ins} < w_{inv} \varepsilon_{ins} / \varepsilon_{si} \qquad (10)$$

This would require a SiO_2 for thickness < 0.5nm for Si and an (Al,Ga)As thickness < 5nm for GaAs. Extrapolating from present L/t_{ins} ratios gives a minimum channel length of about $8w_{inv}$ which, from Table II, are about 11nm, 37nm and 36nm for Si, GaAs and $In_{0.53}Ga_{0.47}As$ respectively. (Of course, using 'reasonable' values for SiO_2, a thickness of 4nm would take us back to the old [5] limit of 100nm and a transconductance of 11S/cm.) Contacting the FET without degradation of g_m would present a severe problem. Contact resistances of $<2\times 10^{-3}$ Ω-cm and specific contact resistivities of $<2\times 10^{-8}$ Ω-cm^2 would be required, about an order of magnitude better than the best values yet reported. Wires would have to carry current densities in excess of 1×10^6 A/cm^2. Operating voltages would typically be about 100mV. An interesting set of relationships exists for this limiting device. The de Broglie wavelength normal to the surface, is of the order of the inversion layer thickness, due to the perpendicular quantization. The condition for equal gate and drain voltages implies that the longitudinal electron momenta in the channel are comparable to the perpendicular, quantized, electron momenta (assuming an isotropic effective mass). Therefore the longitudinal de Broglie wavelength is on the order of the channel thickness which is ~1/8th of the channel length. (For Si, the mass being anisotropic, this ratio is closer to unity.) From this argument one would expect the treatment of the electron as a particle to begin to break down as one approaches the limiting FET size (and even earlier for Si.).

We have conducted an intellectual exercise in stretching semiconductor device concepts to their very limits. Probably, well before we reach such limits logic devices based on completely different operating principles will evolve. Much progress in our understanding of semiconductor physics and analysis techniques will be necessary on our way to getting there. A much better understanding of hetero-interfaces and contacts is needed [30] [31] as these become increasingly important in the scaled-down device. More accurate evaluation of semiconductor band structure will be required, including for the imaginary k vector, which is important in tunneling calculations [32]. Even using the Boltzmann equation, accurate modelling of these devices will present a severe problem. In principle a two-dimensional self-consistent solution of Schrodinger's equation, Poisson's equation and the Boltzmann equation will be necessary. Electron-electron, electron-phonon and phonon-phonon interactions must be better understood. Quantum mechanical effects will become so pervasive in these devices that tomorrow's device designers will have to become 'quantum' mechanics.

The author is indebted to Peter Price and Jeff Tang for many valuable discussions.

REFERENCES

1. R.H. Dennard, F.H. Gaensslen, H.-N. Yu, V.L. Rideout, E. Bassous, and A. LeBlanc, IEEE J. of Sol.-State Circuits, **SC-9**, 256 (1974)
2. G. Baccarini, M.R. Wordeman, and R.H. Dennard, IEEE Trans. Electron Dev. **ED-31**, 452 (1984).
3. B. Hoeneisen and C.A. Mead, Sol.-State Electron., **15**,819 1972.
4. B. Hoeneisen and C.A. Mead, Sol.-State Electron., **15** pp.891-897, 1972.
5. P.M. Solomon and D.D. Tang, ISSCC, Philidelphia P.A., Feb. 1979, Digest p.86.
6. P.M. Solomon,Proc. IEEE, **70**, 489 (1982).
7. J.Barker and D.K. Ferry, Sol. State Electron. **23**, 519 (1980).
8. J.Barker and D.K. Ferry, Sol. State Electron. **23**, 531 (1980).
9. F.H. Gaensslen, V.L. Rideout, E.J. Walker, J.J. Walker, IEEE Trans Electron Dev., **ED-24**, 218 (1977).
10. R.H. Dennard, Physica, **117B**, 39 (1983).
11. F.M. d'Heurle and P.S. Ho, Chap. 8 of "Thin-Films, Interdiffusion and Reactions ", Edited by J.M. Poate, K.N. Tu and J.W. Mayer, John Wiley and sons, Inc., New York, 1978.
12. J.G. Ruch, IEEE Trans. Electron. Dev. **ED-19**, 652 (1972).
13. R.Abram, G.J. Rees and B.L.H. Wilson, Adv. Phys. **27**, 799 (1978).
14. The method of scaling the Poisson equation, at constant voltage, was presented by P.M. Solomon at the 1981 March meeting of the APS, Phoenix Arizona.
15. T.H. Ning, P.W. Cook, R.H. Dennard, C.M. Osburn, S.E. Schuster, and H.N. Yu, IEEE Trans. Electron. Dev., **ED-26**, 346 (1979).
16. Sir Rudolf Peierls, "Some Simple Remarks on the Basis of Transport Theory", Transport Phenomena Lecture Notes in Physics **31**, (1974).
17. P.J. Price, J. Vac.Sci. Technol., **19**, 599 (1981).
18. M. Inoue, M. Inayama, S. Hiyamuzu and Y. Inuishi, J.J. Appl. Phys. Lett., **22**, L213 (1983).
19. W. Fawcett, A.D. Baordman, and S. Swain, J. Phys. Chem. Sol. **31**, 1963 (1970).
20. P.J. Price, Semiconductors andSemimetals **14**, chap. 4 (1979).
21. W.R. Curtice, IEEE Trans. Electron. Dev. **ED-29**, 1942 (1982).
22. R.K. Cook and J. Frey, COMPEL- Int. J. Comput. and Math. Electr. and electron. Eng. (Ireland),**1**, 65 (1982).
23. Y. Awano, K. Tomizawa, and N. Hashizume, IEEE Trans. Electron. Dev. **ED-31**,448 (1984). 24.R.W. Keyes, IEEE J. Solid-State Circuits **SC-12**, 245 (1975).
25. R.H. Dennard, J. Vac. Sci. Technol. **19**,537 (1981).
26. Reviewed in P.M. Solomon and H. Morkoc, IEEE Trans. Electron. Dev. **ED-31**, 1015 (1984).
27. T.S. Ando, A.B. Fowler, and F. Stern, Reviews of Modern Physics **54**, 437 (1982).
28. F. Stern and S. Das Sarma, to be published in Phys. Rev. B, Jul. 1984.
29. F. Stern, Physical Rev. B**5**, 4891, (1972).
30. H. Kroemer, in *Proc. NATO Advances Study Institute on Molecular Beam Epitaxy and Heterostructures,* Erice, Sicily, 1983; ed. by L.L. Chang and M. Ploog, to be published by Martinus Nijhof, The Netherlands.
31. J.L. Freeouf, Surf. Sci. **132**, 223 (1983).
32. G.C. Osburn, J. Vac. Sci. Technol. **17**, 1104 (1980).
33. M.L. Cohen and T.K. Bergstresser, "Phys. Rev. **141**, 789 (1966).

THERMALLY INDUCED TRANSITION METAL CONTAMINATION OF SILICIDE SCHOTTKY BARRIERS ON SILICON

A. Prabhakar and T. C. McGill
California Institute of Technology, Pasadena, California 91125

ABSTRACT

We discuss the results of deep level transient spectroscopy (DLTS), current-voltage (I-V), and Rutherford backscattering spectrometry (RBS) measurements of nickel, palladium, and platinum silicide barriers on n-type silicon which have been annealed at temperatures from 300–800 °C. Reverse-biased leakage currents increase with increasing annealing temperature in all three cases. However, the degradation is almost negligible for the PtSi samples at temperatures which cause the palladium and nickel silicide barriers to become very leaky. For the nickel samples which were still reasonable barriers, DLTS showed no traps. A very small concentration of traps was detected for the high-temperature palladium samples. Significant concentrations of platinum traps were seen in samples of PtSi on silicon annealed at 700 °C and above.

INTRODUCTION

The application of transition metal silicides in LSI and VLSI semiconductor technology is progressing rapidly today.[1] Many of these transition metals form one or more deep level traps in silicon. Indiffusion of transition metals from the silicide layer into the underlying bulk silicon during annealing stages of the processing could cause difficulties. However, the standard techniques — for example, Rutherford backscattering spectrometry (RBS)[2] — presently used to study silicide-silicon structures cannot provide the sensitivity required to detect the small quantity of transition metal contaminants which can poison the underlying silicon ($\geq 10^{11}$ cm^{-3}). Deep level transient spectroscopy (DLTS)[3] is a technique suited for this application. Its sensitivity is $\sim 10^{-4}$ times the shallow dopant concentration, so it is possible to detect in the interesting concentration range.

Here, we present the results of studies of the degradation of platinum, palladium, and nickel silicides on silicon. RBS was used to check the composition of the silicide. Current-voltage (I-V) measurements determined reverse leakage currents. DLTS measurements were used to characterize deep levels in the depletion region in the underlying silicon for the devices with reasonable leakage currents.

EXPERIMENTAL METHOD

The substrates used to fabricate samples for this study were 7-10 Ωcm n-type (100) silicon wafers from Wacker. Wafers were cleaned and then immediately loaded into an ion-pumped vacuum system, where a thin (~ 500 Å) metal film was electron-beam deposited at a pressure less than 3×10^{-7} Torr. During the evaporation, a portion of each wafer was covered by a mechanical mask with 0.75-mm-diam holes to make diodes for the electrical measurements. The wafers were diced, and pieces with broad-area metal coverage were annealed with diodes in a vacuum furnace at pressures below 10^{-6} Torr at temperatures from 300 to 800 °C. Ohmic contacts to the diodes were made by rubbing In-Ga onto the backs of the substrates. Diodes used for DLTS studies were mounted on headers and wire-bonded. The samples with broad-area silicide coverage were used for the backscattering analysis. Further, in some cases in which DLTS measurements could not be made on the silicide-silicon structure due to the degradation of the Schottky barrier, a corresponding broad-area piece was etched in $3HNO_3:1CH_3COOH:0.4HF$ to remove a 5–10 μm layer. Gold dots were then evaporated onto the freshly exposed silicon, and these diodes were prepared as above for DLTS measurements.

Standard RBS spectra were generated by using a beam of 2-MeV $^+He^4$ ions incident on the sample at a few degrees off normal incidence. The detection angle was 170°.

DLTS measurements were made using a Boonton 72BD capacitance meter operating at a frequency of one megahertz. A double-boxcar gating scheme was used to analyze the transients which were produced over the temperature range of ~ 100–320 K. In addition, a 20-MHz bridge was used to analyze faster transients to determine the trap activation energy for comparison to literature values.

RESULTS AND DISCUSSION

RBS spectra were taken for the samples to determine the composition of the thin silicide layer and to check for surface smoothness and interface abruptness. Analysis yields surface silicide compositions consistent with Ni_2Si for 300 °C, NiSi for 400–700 °C, and $NiSi_2$ for 800 °C. Figure 1 shows the results for nickel samples annealed at 600, 700, and 800 °C. The low-energy fall-off of the nickel signal is clearly less abrupt as temperature is increased. This blurring could be due to the interface growing less abrupt, the surface morphology becoming rough, or a combination of the two. Similar spectra for the palladium samples are shown in Fig. 2. RBS indicates that the phase for the 300–800 °C samples is Pd_2Si. Again, we see the palladium signal blurring at its low-energy end. RBS spectra for the various platinum silicide samples have been presented elsewhere.[4] The silicide formed was Pt_2Si for the sample annealed at 300 °C and PtSi for the higher temperature samples. The low-energy fall-off of the Pt signal was abrupt for samples annealed below 800 °C. Significant degradation of the silicide was observed in the sample treated at 800 °C for 180 min.

Reverse-biased leakage currents for nickel, palladium, and platinum silicide Schottky barriers annealed at various temperatures are given in Table 1. The degradation at a given temperature is more severe for palladium than for platinum, and even worse for nickel.

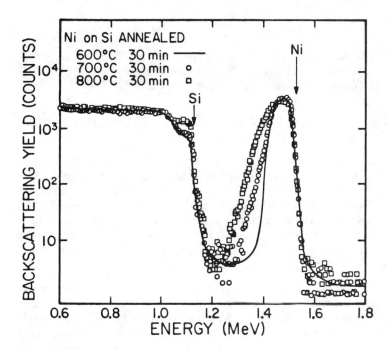

Fig. 1. RBS spectra for nickel-on-silicon structures.

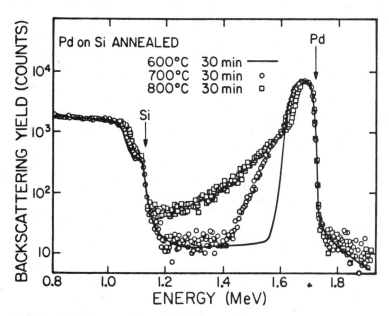

Fig. 2. RBS spectra for palladium-on-silicon structures.

	500 °C	600 °C	700 °C	800 °C
Ni	67 µA	330 µA	2.2 mA	14 mA
Pd	<1 µA	<1 µA	9.1 µA	370 µA
Pt	<1 µA	<1 µA	<1 µA	1 µA

Table 1. Leakage currents at −5 V reverse bias in samples annealed for 30 min.

DLTS spectra were generated with a rate window of 55 s^{-1} for the samples which had sufficiently low reverse leakage currents. The spectra exhibited no traps above a detection limit of ~5×10^{11} traps cm^{-3} in any nickel sample treated below 500 °C. Nickel samples which were treated at temperatures above 500 °C had reverse leakage currents which made DLTS measurements impossible. A piece of nickel material which was annealed at 800 °C was etched and prepared as described above to make DLTS measurements in a region 5–10 µm below the original interface. Again, no traps were detected.

In the case of palladium samples, DLTS spectra exhibited no traps above the detection limit for the samples treated below 700 °C. A barely detectable peak at a temperature of 120 K appeared in the spectrum of the Pd$_2$Si sample annealed at 700 °C (shown in Fig. 3). Although the signal is so small that it is not possible to make a measurement of the trap activation energy, the location of this peak is consistent with the reported activation energy of 0.22 eV from the conduction band.[5] Thus, we believe that the peak represents a small concentration (~ 10^{11} cm^{-3}) of palladium electron traps.

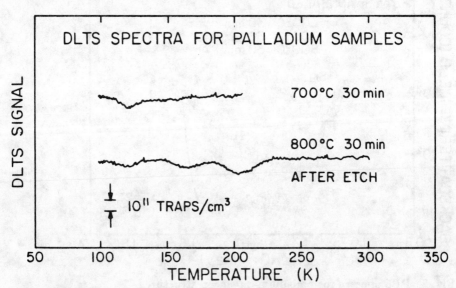

Fig. 3. DLTS spectra of palladium silicide samples taken with boxcar gates at 5 ms and 45 ms for a rate window setting of 55 s^{-1}.

Palladium samples which were treated at temperatures above 700 °C had high reverse leakage currents. As in the case of the high-temperature nickel samples, a piece of 800 °C palladium material was etched to make DLTS measurements below the original interface. The resultant spectrum is shown in Fig. 3. Again, we see a very small peak at 120 K which is probably due to palladium electron traps. In addition, there are minute peaks at 165 and 200 K which were not observed close to the interface in the 700 °C sample. Nor were these peaks seen in the nickel samples which were etched and prepared in exactly the same fashion. These signals may represent the presence of a contaminant (such as iron, a fast diffuser in silicon) which could have been in the palladium charge material used to evaporate the metal onto the silicon substrate. All three peaks correspond to trap concentrations of $\sim 10^{11}$ traps cm^{-3}.

DLTS results for the PtSi samples have been reported in Ref. 4. These samples had very low leakage current even after treatment at 800 °C. No platinum traps were seen in samples annealed below 700 °C, and trap concentrations for the samples treated at temperatures from 700 to 800 °C ranged from $\sim 10^{12}$ to 10^{14} cm^{-3}.

Our RBS studies showed a degradation of the silicide layers for nickel, palladium, and platinum silicides annealed at temperatures above 600 °C. The I-V measurements indicated poor Schottky barrier behavior in nickel samples treated above 500 °C and palladium samples treated above 700 °C. Platinum samples annealed at 800 °C, however, still had reasonable leakage currents. DLTS measurements of the nickel samples with small leakage currents showed no traps. An extremely small number of traps were seen in the 700 and 800 °C palladium samples. Platinum trap concentrations were as high as 10^{14} cm^{-3} in PtSi samples. Thus, we conclude that 700 °C may be regarded as the maximum "safe" temperature at which palladium and platinum silicides may be annealed before transition metal indiffusion begins to degrade the underlying silicon. For our nickel silicide samples, the maximum temperature for good Schottky barrier behavior is only 500 °C.

It is a pleasure to thank Professor M-A. Nicolet for his assistance with this work. This project was supported in part by the Office of Naval Research under Contract No. N00014-84-C-0083.

REFERENCES

[1] F. Mohammadi, Solid State Technol. **24**, 65 (1981).
[2] W.-K. Chu, J. W. Mayer, and M-A. Nicolet, *Backscattering Spectrometry* (Academic Press, New York, 1978).
[3] D. V. Lang, J. Appl. Phys. **45**, 3023 (1974).
[4] A. Prabhakar, T. C. McGill, and M-A. Nicolet, Appl. Phys. Lett. **43**, 1118 (1983).
[5] L. So and S. K. Ghandhi, Solid-State Electron. **20**, 113 (1977).

MOBILITY LIMITATIONS IN VLSI DEVICES
DUE TO THE FIELD-BROADENING EFFECT

Vijay K. Arora[*]
Department of Physics, King Saud University
P. O. Box 2455, Riyadh, Saudi Arabia

ABSTRACT

A first-principles theory, which takes into account the field broadening due to the energy uncertainty $eE\lambda_D$ of an electron of de Broglie wavelength λ_D kept in an electric field E, is presented. Explicit expressions for electron mobility and temperature are derived for deformation potential scattering, and used to explain mobility decrease in Ge, Si, and GaAs, the materials which are extensively used in VLSI device development.

INTRODUCTION

Recent developments in VLSI and VHSIC programs have indicated the ever-increasing importance of high-field effects in limiting the mobility in VLSI devices. It is well known that the mobility in the Ohmic regime is limited by collision broadening $h\tau^{-1}$ dictated by scattering interactions. In the high-field regime, the wave character of an electron plays an important role.[1] Then, the position of an electron is uncertain by λ_D, the de Broglie wavelength. The corresponding energy uncertainty is $eE\hbar_D$ ($\hbar_D \simeq \lambda_D/2\pi$), when an electron is present in an electric field. This gives a finite lifetime $\tau_F \simeq \hbar/eE\hbar_D$ of an electron in a particular state and, hence, the energy of an electronic state is broadened by this field broadening effect ($\hbar/\tau_F \simeq eE\hbar_D$). In low electric fields (Ohmic regime), this field broadening can be assumed to be negligible. Deviations[1] from the linear behavior of electronic current in the Ohmic regime are apparent when $h\tau_F \simeq h\tau^{-1}$ (warm-electron regime), which occurs at $E \simeq 1000$ V/cm in n-Germanium at room temperature. When $h\tau_F^{-1} \gg h\tau^{-1}$ (hot-electron regime), the collision broadening is completely suppressed by the field broadening. This is the regime when velocity saturation is expected and the mobility is solely limited by this field-broadening effect.

These considerations indicate that a radical transformation in our theoretical framework of high-field transport, starting from the

---

[*] Visitor at GTE Laboratories, 40 Sylvan Road, Waltham, MA 02254

fundamental principles of quantum and statistical mechanics, is needed. To keep calculations manageable and expose clearly the importance of field-broadening effect, the size limitations on devices due to the quantum confinement effects[2] are ignored here.

THEORY

The linearized distribution function, for acoustic-phonon scattering, is well known to be:[3]

$$f = f_0 + (e\vec{p} \cdot \vec{E}/m^*)\tau(\varepsilon) \partial f_0/\partial \varepsilon, \qquad (1)$$

with

$$f_0 = [\exp\{(\varepsilon - \zeta_0)/k_B T\} + 1]^{-1} \qquad (2)$$

$$\approx \exp[(\zeta_0 - \varepsilon)/k_B T], \text{ (nondegenerate)}. \qquad (3)$$

Here, $\vec{p} = \hbar \vec{k}$ is the carrier momentum vector, $\vec{E}$ is the electric field, m^* is the mass of an electron in the parabolic band model, ε is the carrier's kinetic energy, and T is the temperature. The momentum relaxation time $\tau(\varepsilon)$ is given by[3]

$$\tau(\varepsilon) = [\pi \hbar^4 \rho_d u^2 / (2^{1/2} E_1^2 m^{*3/2} k_B T)] \varepsilon^{-1/2}, \qquad (4)$$

$$\equiv l_a/v, \qquad (5)$$

with

$$l_a = \pi \hbar^4 \rho_d u^2 / m^{*2} E_1^2 k_B T, \qquad (6)$$

where ρ_d is the crystal density, u the longitudinal speed of sound, E_1 the deformation potential constant, and v the thermal velocity of an electron. The Fermi energy (chemical potential) ζ_0 for nondegenerate semiconductors is obtained as

$$\zeta_0 = k_B T \ln(n_e/N_c), \qquad (7)$$

with

$$N_c = 2(2\pi m^* k_B T/h^2)^{3/2}, \qquad (8)$$

where n_e is the number of electrons per unit volume.

The equilibrium distribution function of Eq.(2) will change significantly in the presence of an electric field, and chemical potential ζ_0 transforms to an electrochemical potential ζ. Then, it may not be possible to write the distribution function in the form of Eq.(1). Following the procedure adopted by Zukotynski and Howlett[4], which determines the proper selection of states from the principle of equal a priori probabilities with an added constraint of steady state current, a distribution function with an undetermined vector Lagrange's multiplier is obtained. The value of this multiplier can easily be obtained by expanding the distribution function so obtained in the linear approximation, and comparing this with Eq.(1) obtained from the Boltzmann transport equation. The final result for the nonlinear distribution function is

$$f(\vec{k}) = [\exp\{(\varepsilon-\zeta)/k_B T - \delta\cos\theta\} + 1]^{-1} \tag{9}$$

$$\simeq \exp[(\zeta-\varepsilon)/k_B T + \delta\cos\theta], \text{ (nondegenerate)} \tag{10}$$

with

$$\zeta = \zeta_0 + k_B T \ln(\delta/\sinh\delta), \text{ (nondegenerate)}, \tag{11}$$

$$\delta = eEl_a/k_B T, \tag{12}$$

where θ is the angle between the electric field vector and the momentum vector. The physical interpretation of Eq.(9) is very simple. In addition to the thermal kinetic energy of an electron, the distribution function also contains a potential energy $V = e\vec{E}\cdot\vec{l}_a$ of an electron confined to a mean-free-path l_a of an electron. This lowers the electrochemical potential of an electron, tending towards lower degeneracy of the system. The transition from Ohmic to non-Ohmic regime can be assumed to take place at $E = E^*$ when $\delta=1$ (which is equivalent to the condition $h\tau_F^{-1} \simeq h\tau^{-1}$ stated earlier):

$$E^* = k_B T/e\, l_a \,. \tag{13}$$

Eq.(13) indicates that the nonlinear effects are more important at low temperatures and in high mobility materials (infrequent collisions). Since $l_a \sim T^{-1}$, $E^* \sim T^2$ for acoustic-phonon deformation potential scattering.

The mobility μ, using the distribution function of Eq.(9), is easily obtained as

$$\mu = \mu_0(3/\delta)\, L(\delta), \tag{14}$$

with

$$\mu_0 = 4e\, l_a/3\, (2\pi m^* k_B T)^{1/2}. \tag{15}$$

Eq.(14) gives $\mu = \mu_0$ in the ohmic limit ($\delta \to 0$), $\mu = \mu_0(1-\delta^2/15)$ in the warm-electron limit ($\delta \ll 1$), and $\mu = 3\mu_0/\delta$ in the hot-electron limit ($\delta \gg 1$). Thus, the saturation drift velocity of an electron is obtained for acoustic-phonon scattering. To include the temperature-dependence of the saturation velocity v_{sat}, we obtain from the high-field limit of Eq.(14) multiplied by an electric field E, an expression

$$v_{sat} = (8 k_B T/\pi m^*)^{1/2} (1-\delta^{-1}), \qquad (16)$$

which is almost independent of scattering parameters and is equivalent to the expression obtained if τ^{-1} is replaced by τ_F^{-1} in the Boltzmann transport framework.

DISCUSSION AND APPLICATIONS

The results obtained above are in direct contrast to those obtained by earlier theories, most prominent of which are expansion-method theories. A comprehensive review of these theories is given by Nag[5], who has also compounded the experimental data on elemental semiconductors. The major difference in the outcome of these theories from Eq.(14) or (15) is that the velocity saturation cannot be obtained by the acoustic phonon scattering. The critical field for the onset of nonlinearity given by Shockley[6] is $E^*_s = 1.51$ c/μ_0, which contrasts with $E^* = 2v_{rms}/3 \; \pi^{1/2} \mu_0$ obtained from Eq.(13), where $v_{rms} = (2k_B T/m^*)^{1/2}$ is the random electron velocity. The acoustic-phonon velocity in Shockley's condition is thus replaced by the random thermal velocity in Eq.(13).

In Fig. 1, we show a general plot of the relative mobility μ/μ_0 as a function of normalized electric field $\delta = E/E^*$ as obtained from Eq.(14). The electric field $E_{1/2}$ at which mobility falls to half its Ohmic value $E_{1/2} = 4.78 \; E^*$. In n-Germanium, at 300 K, $E_{1/2} = 2.8$ kV/cm,[5] giving $E^* = 0.59$ kV/cm. With this value of E^*, the relative mobilities obtained from Eq.(14), $\mu/\mu_0 = 0.85, 0.62,$

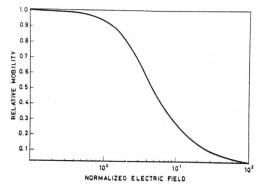

Fig. 1 Relative Mobility Versus Normalized Electric Field.

0.47, and 0.38, are in excellent agreement with the median experimental values[5] $\mu/\mu_0 = 0.82$, 0.62, 0.46 and 0.38 at $E = 1, 2, 3, 4$ kV/cm, respectively. Similarly, at 77 K, with $E^* = 0.11$ kV/cm corresponding to $E_{1/2} = 0.52$ kV/cm, the theoretical values of 0.51, 0.29, 0.16, and 0.10 compare well with $\mu/\mu_0 = 0.52$, 0.29, 0.16, and 0.10 obtained experimentally at $E = 0.5, 1, 2, 3$ kV/cm, respectively (see Zucker in Ref. 5).

Silicon is the most extensively used material in VLSI devices. The room-temperature experimental values of $\mu/\mu_0 = 0.98, 0.87, 0.79, 0.72, 0.65$ (see Davies and Goslin in Ref. 5) are in close agreement with $\mu/\mu_0 = 0.97, 0.90, 0.81, 0.71$, and 0.63 at $E = 1, 2, 3, 4, 5$ kV/cm, respectively, if $E^* = 1.5$ kV/cm is taken.

In GaAs, below the threshold electric field for negative differential conductivity, the nonlinear effects are also observed.[7] At temperatures of $T = 10, 77, 100$, and 120 K, the relative mobilities obtained at $E = 200$ V/cm are $\mu/\mu_0 = 0.27, 0.38, 0.57, 0.74$, respectively, in a superlattice of 50 nm thickness. These correspond to E^* values of 20, 29.4, 51.2, and 80 V/cm, respectively. This gives $E^* \sim T^{2.2}$ for intermediate temperatures of $T = 77, 100, 120$ K, which is in approximate agreement with $E^* \sim T^2$ predicted by Eq.(13). Similarly, for 25 nm layers, $\mu/\mu_0 = 0.22, 0.41, 0.53$, and 0.83 obtained at $T = 10, 77, 100, 170$ K give $E^* = 16, 33, 46, 111$ V/cm, respectively, resulting in $E^* \sim T^{1.5}$ at intermediate temperatures of $T = 77$ K, 100 K, 170 K. In thin devices, the quantum confinement effects[2] reduce the effective mean free path by a factor of $d/\lambda_D (d \ll \lambda_D)$, where d is the layer thickness, and $\lambda_D \sim T^{-1/2}$. Therefore, it is not surprising to find $E^* \sim T^{-1.5}$, as compared to $E^* \sim T^{-2}$ for devices retaining bulk character for electrons.

The simplified model considered above has not included the details of anisotropic nature of band structure in many-valley semiconductors, nor the inter-valley transfer of electrons which yield negative differential conductivity. Neither have we considered other dominant scattering mechanisms. Nevertheless, we hope that the result presented above will form a nucleus for more advanced studies of high field transport in VLSI devices, thereby contributing effectively to the modelling process of these devices.

ACKNOWLEDGEMENTS

The author is grateful to Drs. Scott Norman and Johnson Lee for giving access to their facilities during his visit to GTE Laboratories.

REFERENCES

1. V. K. Arora, J. Appl. Phys. 54, 824(1983).
2. V. K. Arora and F. G. Awad, Phys. Rev.B 23, 5570(1981).
3. B. R. Nag, Electron Transport in Compound Semiconductors, (Springer-Verlag, Berlin, 1980).
4. S. Zukotynski and W. Howlett, Sol. State Electron. 21, 35(1978).
5. B. R. Nag, Sol. State Electron. 10, 385(1967).
6. W. Shockley, Bell Syst. Tech. J. 30, 990(1951).
7. T. J. Drummond, M. Keever, W. Kopp, H. Morkoc, K. Hess, B. G. Streetman, and A. Y. Cho, Electron. Lett. 17, 545(1981).

SCALING PROBLEMS AND QUANTUM LIMITS IN INTEGRATED CIRCUIT MINIATURIZATION

H.-U.Habermeier

Max-Planck-Institut für Festkörperforschung,Stuttgart,FRG

ABSTRACT

The trends in integrated circuit technology are directed towards an enhancement of packing density and the reduction of transistor feature size by means of the usual scaling concept.Currently, the state of the art in miniaturization is restricted by the technological feasibility of lithography and pattern transfer rather than by limits associated with basic solid state physics.In the future,however, as sub o.1 μm design rules become available, the usual scaling concept is no longer valid and physical effects due to small dimensions may play a more important role in device performance. In this paper some conceptual problems are discussed and quantum mechanical limits on device performance will be treated.

INTRODUCTION

The integrated circuit industry has enjoyed tremendous growth and success during its first two decades,mainly based on advances in microfabrication technology and improvements of material properties. The trends in the IC technology are still directed to an enhancement of packing density and thus to the reduction of the feature size of individual transistors on a chip.Associated with these efforts to enhance the integration density are attempts to keep the total energy dissipation in a chip below the limit of 1 W/cm^2 chip area.Currently, the state of the art in miniaturization is limited by the technological feasibility of lithography,pattern transfer and packaging problems rather than by fundamental physical laws and problems due to material properties.However,as the "transition" from VLSI to ULSI will be realized and sub 0.1 μm design rules will be envisaged,the usual scaling concept may be no longer valid and limits based on conceptual problems and quantum effects may play a dominant role.

In integrated circuit miniaturization the designer follows the basic ideas of the scaling concept as introduced by Dennard et al.[1] The procedure starts from a long channel MOSFET,scales down all device dimensions and voltages by the same factor K and enhances the doping concentration by K. After that, the basic physics of the new device is the same as that of the original one.This simple concept of constant field scaling can be refined to nonconstant field scaling concepts in which different scaling factors are defined for voltages, the gate oxide, doping level and all horizontal and vertical dimensions^2. Nevertheless, as soon as device dimensions become smaller than characteristic dimensions, methods used to describe the bulk behaviour become inapplicable and properties based on the physics of particles have to be considered. In the second section of this paper some conceptual problems due to the reduced size of transistors and interconnection lines are listed and quantum mechanical limitations of device performance are discussed.

CONCEPTUAL LIMITATIONS OF SCALING

The basic concept of a long channel MOS (or bipolar) transistor and thus of scaled devices for VLSI and VHSIC-applications is described within the framework of conventional solid state physics in three dimensional systems as developed during the last 50 years. However, as device dimensions are reduced below characteristic distances such as the de Broglie wavelength, electron mean free path, mean free path for inelastic scattering and grain size of polycrystalline layers mechanical and electrical properties change and the principles used to describe the bulk behaviour become more and more inapplicable with decreasing feature size. In table I some of those properties and methods are listed, the table respresents a selection only and is far from exhausting.

Table I: Selection of concepts becoming inapplicable with increasing miniaturization.

Bulk behaviour	"Particle" behaviour
Lines ("wires") are mechanically stable	Instability of fine lines ("fine wires") due to grain boundary grooving
Metallic interconnection lines behave metallic	Interconnection lines behave nonmetallic at low temperatures
Distribution of doping atoms can be treated randomly	Clustering effects of doping atoms possible
Energy levels are continuous	Energy levels are discrete
Elecron transport described by ordinary Boltzmann equation	Time scale reduced, i.e. collision time ∾ transit time

To take the full advantage of small dimension devices,the interconnection lines to wire the device into VLSI circuits should be reduced in size ,too,and scaled by K^2. Without treating electrical implications like RC- time constants and switching speeds,there are some mechanical aspects involved in scaling of interconnection lines. Due to the large surface to volume ratios in fine lines,point defects, dislocations and grain boundaries can move easily to the free surface and generate elastic strains in the individual grains.If the geometric pattern generated by microfabrication techniques e.g. is rectamgular and the defect distribution is somehow asymmetric, grain boundaries oriented parallel to the fine line are likely to be eliminated in contrast to those perpendicular to the line.The result is a modification of the line shape from rectangular to a bamboo-like structure. This resulting bamboo structure has some implications concerning transport properties.Electromigration, e.g. occurring in"wide" thin polycrystalline films lead to a degradation of electronic devices. This effect is believed to be due to grain boundary diffusion. In the bamboo structure the dominant mechanisms for mass transport are interface and surface diffusion and thus,the problems of electromigration are reduced.This desirable property,however, may be overcompensated by the mechanical instability of the structure due to grain boundary grooving[4] and surface roughening[5].

Another limitation of miniaturization may be seen in the distribution of impurity atoms (dopants and other impurity atoms).In long channel devices a continuous distribution of doping atoms is assumed with spatial variations occurring on a macroscopic scale. As the device dimensions are reduced,the assumption of a "jellium"-like distribution is less realistic and fluctuationsof the number of impurity atoms within a given volume determine the device performance.These fluctuations affect the actual values for the breakdown voltage (due to field emission) and the punch -through voltage (two pn- junctions close together).In the case of bipolar devices Hoeneisen et al.[6] give an estimation for the lower limit of the base width of 0.05 μm for silicon according to punch through effects.

From the point of view of physics,the basic changes in sub-0.1 μm devices occur in the elctron transport properties. Conventional treatments of electron transport apply, when the carrier mean free path,l, is much smaller than the sample length,L.In this case the electron transport is dominated by collisions and the transport properties are governed by the ordinary Boltzmann equation in the relaxation time approximation.At $L \sim l$,the transport becomes more and more ballistic.Such size effects in electron transport have been treated in the case of applied voltages larger than the thermal voltage by Shur and Eastman[7].In this case carrier injection and plasma effects lead to a periodic variation of the electron velocity, concentration,field and potential within the sample.When a voltage smaller than thermal voltage is applied to the sample,the low field mobility decreases with finite sample length by a factor of $(1-\exp(-3L/2l))$[8]. In both cases long channel transport properties cannot just be scaled down.

Another example for the breakdown of the scaling concept is the electron transport in one and two dimensional metallic systems (interconnection lines) at cryogenic temperatures. Due to the reduced dimension quantum corrections to the electrical resistance, ΔR, occur and such systems behave as insulators[9] as the temperature approaches zero. In the case of one dimensional disordered materials (thin wires) the temperature dependence of these quantum corrections are described as $\Delta R_1 \sim T^{-1/2}$ as shown by Giordano[10] as well as Chaudhari and Habermeier[11]. In two dimensional systems like thin films a logarithmic increase $\Delta R \sim \ln(T/T_o)$ is observed[12]. As shown in Fig.1 this logarithmic increase of the resistance has been observed not only in disordered materials but also in single crystalline films[13].

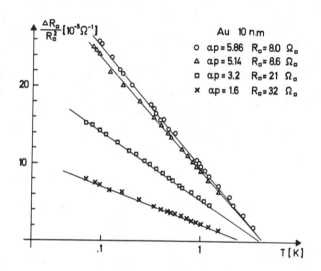

Fig.1: Temperature dependence of the sheet resistance of single crystal gold films.

QUANTUM LIMITS OF DEVICE PERFORMANCE

The limits treated in this section are beyond the scope of the present state of the art and may be irrelevant to the current technology, but looking ahead at problems and prospects of the possibility of "chemical" computers based on electronic devices with molecular dimensions[14], such a quantum jump requires some considerations concerning the basic physics of logical devices[15].

The parameter to be optimized in logic devices is the power-delay-time product with the constraint of a permitted error rate. If p_L is the average power dissipated at the maximum clock rate and t_D is the average delay time introduced by a gate to make its logic decision, the product $p_L t_D$ is the number of interest. In an abstract logic gate model Landauer[16] showed that the minimum energy dissipation in a "restore to one" operation depends on the temperature of the switching element, only and is $kT\ln2$ ($=4\times 10^{-6}$ fJ at room temperature). This holds as long as the relaxation time τ is much smaller than the switching time t_s. Classically, τ can be reduced arbitrarily to achieve fast switching and simultaneously, the energy difference, ΔE, can be enhanced arbitrarily to reduce the error probability. However, if we treat the logic gate model as a quantum system, Heisenberg's uncertainty principle imposes a quantum limit of the order of $E \gtrsim \hbar/t_s$ for the energy dissipated per logic operation. In terms of clock frequencies $f_c = t_s^{-1}$ the quantum limit exceeds the thermodynamic limit at

$$f_c = \frac{kT}{\hbar}\ln2 \qquad (1)$$

At room temperature, this crossover frequency occurs at 27THz. Quantum effects like level broadening and quantum limits on the error probability, however, will reduce this crossover frequency dramatically as shown by Bates[17]. In a hypothetical system e.g. with a functional throughput rate of 10^{18} gates·Hz (i.e. 10^6 gates at 1 THz) quantum limits will occur for clock frequencies larger than 75 GHz if an error rate of $(10^9 hr)^{-1}$ is assumed.

CONCLUSION

Accompanying integrated circuit miniaturization beyond the 0.1 μm limit is the breakdown of the classical scaling concept due to the increasing importance of quantum effects occurring in small systems. In addition to those more "practical" problems, the quantum nature of a two level model for logic switching has tradeoffs concerning energy dissipation, switching time and error probability. To estimate the principal capability of logic systems all these topics have to be investigated in detail. They span a new broad field of activitites in basic as well as applied research.

ACKNOWLEDGEMENTS

This work was supported by the "Bundesministerium für Forschung und Technologie", Bonn, West Germany, under contract number NT 2582.

REFERENCES

1. R.H.Dennard,T.H.Gaensslen,H.N.Yu,V.L.Rideout,E.Bassous and A.R. LeBlanc,IEEE SC-9,256 (1979)
2. P.K.Chatterjee,W.R.Hunter,T.C.Holloway and Y.T.Lin,IEEE EDL-1, 22o (1981)
3. A.L.Ruoff and K.-S. Chan,"VLSI Electronics " (N.G.Einspruch ed.) Vol.V,p.343, Academic Press,N.Y. 1982
4. P. Chaudhari, IBM Research Report Nr. RC 7413 (1978)

5. R.B.Laibowitz,A.Broers,J.Viggiano and J.Cuomo,Bull.Am.Phys.Soc. $\underline{23}$,357 (1977)
6. B.Hoeneisen and C.A.Mead, Solid State Electron. $\underline{15}$, 819 (1972)
7. M.S.Shur and L.F.Eastman,IEEE ED $\underline{26}$, 1677 (1979)
8. A.A.Kastalski and M.S. Shur, Solid State Comm. $\underline{29}$,715 (1981)
9. D.J. Thouless,Phys. Rev.Lett.$\underline{39}$,1167 (1977)
1o. N.Giordano, Phys.Rev.$\underline{B22}$,5635 (198o)
11. P.Chaudhari and H.-U. Habermeier, Phys. Rev. Lett.$\underline{44}$,4o (198o)
12. G.J.Dolan and D.D.Osheroff,Phys.Rev.Lett.$\underline{43}$,721 (1979)
13. H.-U.Habermeier and P.Chaudhari, Proceedings of the Conference on Localization,Interaction and Transport Phenomena in Impure Metals,Braunschweig,1984, in the press.
14. F.L.Carter, NRL Memorandum Report 4717 (1982)
15. R.W.Keyes, Proc.IEEE $\underline{63}$,74o (1975)
16. R. Landauer, IBM J.Res. Dev. $\underline{5}$,183 (1961)
17. R.T.Bate, "VLSI Electronics " (N.G.Einspruch ed.) Vol.V,p.359, Academic Press,N.Y. 1982.

Part VIII. Materials Processing

The Application of Silicon Molecular Beam Epitaxy to VLSI

J. C. Bean

AT&T Bell Laboratories
Murray Hill, New Jersey 07974

ABSTRACT

Silicon molecular beam epitaxy is reviewed emphasizing opportunities for application in VLSI. These include: 1) exploitation of MBE control and uniformity in device scaling and/or process refinement; 2) use of mathematically arbitrary dopant profiles in optimized device structures; 3) MBE heteroepitaxy of crystalline insulators, metals and semiconductors; 4) combination of doping and heteroepitaxial capabilities in heterojunction devices. In each of these areas examples are provided from the recent literature.

This paper will focus on the application of Si-MBE[1-4] to VLSI. Opportunities can currently be identified in several areas, including the following: 1) reduction of silicon epitaxial layer thickness with enhanced control and uniformity facilitating device scaling and/or yield enhancement; 2) application of hyperabrupt doping profiles in conventional (e.g., bipolar or microwave) or unconventional (e.g., triangular barrier or N-I-P-I superlattice) device structures; 3) heteroepitaxial integration of silicon with crystalline insulators, metals and non-silicon semiconductors; 4) the combination of hyperabrupt doping profiles with heteroepitaxy to produce two-dimensional transport effects (e.g., modulation doping). Following a brief summary of Si-MBE capabilities, the above applications will be illustrated by examples from the recent literature.

As applied to silicon, molecular beam epitaxy is a relatively straightforward process. Chambers include e-beam evaporation sources, shutters, deposition sensors, heaters and moving substrates arranged in a manner very similar to that of a conventional metalization system.[5] The principal differences stem from the need to prepare and maintain atomically clean, crystallographically ordered surfaces. Samples can be cleaned using a conventional ion milling gun if an ultrahigh vacuum environment is provided. This dictates the use of baked, metal-sealed chambers with sample loading interlocks. Given this environment, once a clean substrate is prepared epitaxial growth may proceed at temperatures as low as 450-800°C. At such temperatures diffusion and autodoping effects are completely absent, and deposition literally can be tailored on an atomic layer by layer basis. Although one might expect inferior epitaxy at such low temperatures, Si-MBE layers have attained a quality comparable to CVD films: zero stacking faults, line defect densities down to $100/cm^2$, bulk mobilities and minority carrier lifetimes ~100 μsec, DLTS trap densities at the noise level and oxide interface state densities ~$5 \times 10^{10}/cm^2$-V.

In current Si-MBE research equipment, layer thickness and uniformity can be controlled to within 1-3% using 3-4" wafers for epitaxial layers of 10Å–10 μm. Assuming wafer size and throughput can be scaled economically and that no unexpected process-induced problems are encountered, these capabilities suggest a near-term application of Si-MBE: that of conventional device scaling and/or yield enhancement. An example of this process is offered by the recent work of Kasper and coworkers[6] at Telefunken. They applied Si-MBE to a frequency divider circuit (Fig. 1) involving linear preamplifiers and master-slave flip-flops in emitter coupled logic (ECL). The original circuit incorporated a 2.5 μm CVD layer and yielded a maximum operating frequency of 900 MHz. By substituting a thinner MBE layer grown at temperatures where substrate out-diffusion was absent, the effective epi thickness was reduced to 0.4 μm.

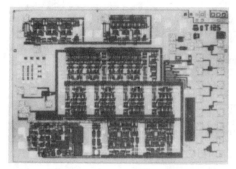

Fig. 1 Telefunken frequency divider circuit using emitter coupled logic. Substitution of MBE for CVD resulted in an increase of operating frequency from 900 MHz to 2.5 GHz with comparable device yields (Ref. 6 by permission of author).

500 μm

This facilitated scaling of the ECL structure and produced a three-fold enhancement of operating frequency to 2.5 GHz. As significant, these layers (grown in a home built research system) produced an overall circuit yield of ~80%, a figure comparable to that achieved with the much more mature CVD process. With refined, commercially built Si-MBE equipment (now becoming available) still better control and quality can be expected and a yield advantage may be realized. As to the question of economic wafer-size and throughput scaling, it should be noted that several research-equipment vendor collaborations are actively addressing this problem. In our laboratory, by mid 1985, we hope to bring on-line a machine capable of simultaneous deposition on multiple wafers up to 200 mm in diameter, with a throughput comparable to CVD reactors.

With proper equipment, dopant atom incorporation may be controlled as readily as the primary silicon molecular beam flux. Further, once dopant atoms are incorporated they are effectively frozen in place given the negligible diffusivities at common MBE growth temperatures. This capability suggests a second VLSI application of Si-MBE: to synthesize mathematically arbitrary doping profiles in conventional or unconventional device structures. "Mathematically arbitrary" may mean a hyperabrupt high-low profile with a $\sim 10^3$ change in doping level in less than 100Å, or a profile incorporating multiple narrow dopant pulses, or the $X^{-3/2}$ profile called for in an ideal varactor diode. Or it may mean simply producing a truly intrinsic few thousand angstrom thick layer on a heavily doped substrate for subsequent MOS processing. All of the above have been successfully grown using Si-MBE.[1,7] A dramatic example of this capability is provided by the TUNNETT doping profile[8] of Fig. 2. The ability to tailor dopant profiles means that, assuming subsequent process induced diffusion can be minimized, the device designer is no longer constrained by Gaussian or error function profiles but can instead implement fully optimized designs. This presents obvious application opportunities in bipolar and microwave devices and emerging possibilities in triangular barrier[9,10] or N-I-P-I superlattice[11] structures where dopant pulses are used to modify anything from carrier conduction to fundamental band structure.

Because Si-MBE is an inherently low-temperature process, not only is dopant motion minimized but all atomic displacements are reduced. It is thus natural that MBE be applied to heteroepitaxial growth where high-temperature processing would otherwise induce interdiffusion, grown-in strain, cross-doping, islanded growth or other degrading effects. To date, MBE has been used successfully for heteroepitaxial growth on and of crystalline insulators (Al_2O_3, $MgAl_2O_4^{12}$, ZrO_2^{13} and CaF_2^{14}), metals ($CoSi_2$ and $NiSi_2^{15-18}$), and nonsilicon semiconductors (GaP^{19} and Ge^{20-21}). Further, unlike CVD, these heteroepitaxial capabilities generally require no fundamental modification of the MBE system but simply the addition of appropriately charged sources. The charges in these sources can be readily changed, and indeed in our lab we have switched between fully calibrated crystalline $NiSi_2$ to Ge_xSi_{1-x} growth in only one day. The body of MBE heteroepitaxial growth literature is too large to summarize in this paper, but two examples will provide a flavor of the possibilities.

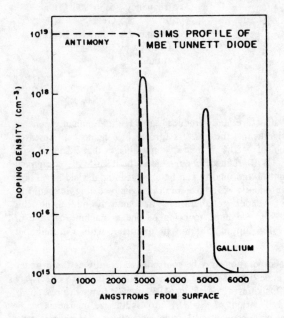

Fig. 2   SIMS profile of dopant distribution in Si-MBE layer grown for TUNNETT diode application (after Ref. 8 by permission of author).

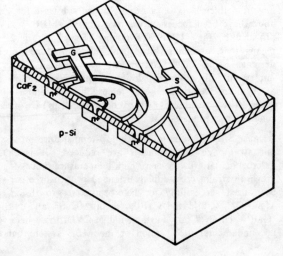

Fig. 3   Schematic diagram of Metal-Epitaxial-Insulator-Semiconductor-Field-Effect-Transistor (MEISFET) fabricated by MBE (Ref. 22 by permission of author).

CaF$_2$ is a relatively unique material in that it readily evaporates as a stoichiometric molecule, it has a cubic crystalline structure similar to and closely lattice matched with silicon, and it has good insulating properties. Its use in Si-MBE was first demonstrated by Ishiwara[14] and has subsequently been pursued in a number of laboratories. Very high quality CaF$_2$/Si heterostructures have been produced and these heterostructures have been overgrown with good (but not yet perfect) silicon epitaxy. If the silicon overgrowth can be perfected this materials system will have obvious applications in the heavily investigated SOI area. Perhaps not so obvious is the possibility that epitaxial CaF$_2$ alone may have an important application. This is because after decades of investigation SiO$_2$ continues to be one of the more difficult materials in VLSI. Its quality synthesis requires high temperature steps in processes increasingly driven to lower temperatures. The Si/SiO$_2$ interface retains a finite defect density variously attributed to dangling bonds, lack of stoichiometry, or impurities. In contrast, not only can crystalline CaF$_2$ be grown at low-temperatures, but given the epitaxial nature of the CaF$_2$/Si interface (where atomic coordination is carefully maintained) a much lower interface state density can be expected. This possibility has been investigated by a Brown University-Bell Labs group using the MEISFET[22] (Metal-Epitaxial-Insulator-Semiconductor-FET) device shown in Fig. 3. In this first attempt at an FET structure, interface state densities as low as $7 \times 10^{10}/\text{cm}^2$–V have already been measured and lower values are anticipated.

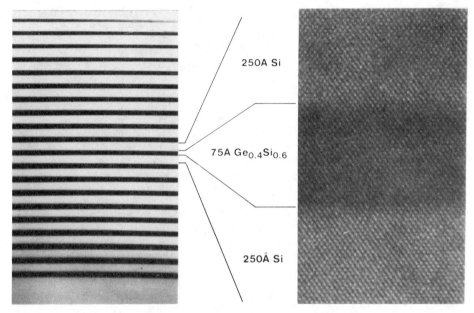

Fig. 4   TEM cross sections of Ge$_x$Si$_{1-x}$/Si strained-layer-superlattice at low and high magnification. Note absence of misfit dislocations in situation where relaxed layers have ~2% lattice mismatch (Ref. 3).

A second example of heteroepitaxial possibilities is provided by the Ge$_x$Si$_{1-x}$/Si system. Given the 4.2% lattice mismatch between Ge and Si the conventional wisdom would suggest that large defect densities are unavoidable. In fact, using low temperature MBE (with careful control of substrate preparation and growth conditions) not only have defect-free heterostructures been grown but as shown in Fig. 4, superlattices with twenty or more periods may be synthesized.[3,20,21] Growth occurs by what has become known as strained-layer-epitaxy.[23] In the

Ge_xSi_{1-x}/Si system the alloy layers compress in the plane of growth to match the bulk Si lattice parameter. This misfit dislocation free compression can be maintained for 3/4 μm in dilute alloys ($Ge_{15}Si_{85}$) or to lesser thicknesses in more concentrated alloys (e.g., 100Å with $Ge_{50}Si_{50}$). The ability to combine silicon with another semiconductor (with its differing electronic, optical and mechanical properties) opens the door to basic and applied possibilities which have, until now, been the exclusive preserve of the III-V and II-VI semiconductors.

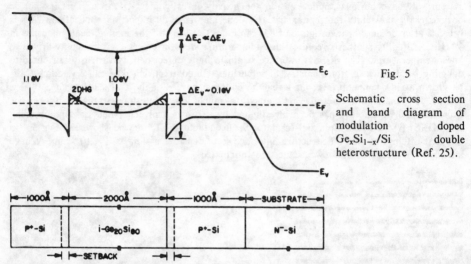

Fig. 5

Schematic cross section and band diagram of modulation doped Ge_xSi_{1-x}/Si double heterostructure (Ref. 25).

A fourth possible VLSI application of Si-MBE is apparent in the combination of the above described hyperabrupt doping and heteroepitaxial growth capabilities. Exploiting these degrees of freedom, conductivity and band structure may be decoupled and a wide variety of new device structures becomes attainable. One of the better known possibilities is that of the heterojunction bipolar transistor[24] where band structure is used to produce high current gain removing constraints on dopant profiles, and permitting independent optimization of parameters such as base series resistance. More radical opportunities are offered by exploitation of two dimensional transport effects such as modulation doping. Indeed, modulation doping has recently been achieved in silicon based materials using the MBE structure shown in Fig. 5. In this device,[25] carriers generated at the acceptor atoms in the large bandgap Si fall into the undoped narrower bandgap $Ge_{20}Si_{80}$ well. This segregation is confirmed by the absence of low-temperature carrier freeze out, by the observation of Shubnikov de Haas oscillations in the low-temperature magnetoresistance, and most importantly by the observation of enhanced carrier mobilities. In early experiments low-temperature Hall hole mobilities have approached 5000 cm^2/V-sec, a value equal to or exceeding the best P-MOS channel mobilities. With optimization, hole mobilities approaching those of π–Si may be achieved and experiments with other alloy compositions should yield the complementary N-type structure.

To summarize, a wide range of VLSI applications are now indicated for silicon molecular beam epitaxy. These include both conventional, near-term, opportunities in the areas of circuit scaling and yield enhancement and the more radical, physics-laden possibilities of two dimensional transport and "bandgap engeneering." Concrete demonstrations now exist in all of these areas. These demonstrations, combined with ongoing work on process scaling at a number of international laboratories, suggest Si-MBE application not only in specialty devices but in broad areas of commercial electronics.

REFERENCES

(1) Review: Growth of Doped Silicon Layers by Molecular Beam Epitaxy, J. C. Bean, Ch. 4, Impurity Doping Processes in Silicon, F. F. Y. Wang, ed., North Holland, Amsterdam (1981).

(2) Review: Silicon Molecular Beam Epitaxy as a VLSI Processing Technique, Proc. Int. Elect. Device Mtg., 1981, p 6.

(3) Review: Silicon MBE: From Strained-Layer Epitaxy to Device Application, J. C. Bean, Proc. 6th Amer. Conf. on Crystal Growth, to be published J. Crystal Growth early 1985.

(4) Bibliography: J. C. Bean and S. R. McAfee, journal De Physique, Colloque C5,. Supplement 12, 153 (1982).

(5) Silicon MBE apparatus for Uniform High-Rate Deposition on Standard Format Wafers J. C. Bean and E. A. Sadowski, J. Vac. Sci. Technol. 20, 137 (1982).

(6) Application of Si-MBE for Integrated Circuits, E. Kasper and E. Worner, VLSI Science and Technology/1984, p. 429; K. E. Bean and G. A. Rozgonyi, Eds., Electrochemical Soc., Pennington, NJ (1984).

(7) Space Charge Behavior of Thin-MOS Diodes with MBE-Grown Silicon Films, U. Lieneweg and J. C. Bean, to be published, Sol. St. Electron.

(8) Sharp Profiles with High and Low Doping Levels in Silicon Grown by Molecular Beam Epitaxy. S. S. Iyer, R. A. Metzger and F. G. Allen, J. Appl. Phys. 52, 5608 (1981).

(9) Rectifying Variable Planer-Doped-Barrier Structures in GaAs, R. J. Malik, K. Board, L. F. Eastman, L. E. C. Wood, T. R. AuCoin, and R. L. Ross, Inst. Phys. Conf. Ser. No. 56 Ch. 9, p. 691 (1981).

(10) Silicon Triangular Barrier Diodes by MBE using Solid Phase Epitaxial Regrowth, D. C. Streit and F. G. Allen, IEEE Elec. Dev. Lett. EDL-5, 254 (1984).

(11) Semiconductor Superlattice — A New Material for Research and Applications, Gottfired H. Dohler, Physica Scripta, Vol. 24, 430 (1981).

(12) Growth of Silicon Films on Sapphire and Spinel by Molecular Beam Epitaxy. J. C. Bean, Appl. Phys. Lett. 36, 741 (1980) and references therein.

(13) Characterization of the Initial Growth of Si on Cubic Stabilized Zirconia, V. A. Loebs, T. W. Haas and J. S. Solomon, J. Vac. Sci. Technol. Al, 596 (1983).

(14) Silicon/Insulator Heteroepitaxial Structures Formed by Vacuum Deposition of CaF_2 and Si, H. Ishiwara and T. Asano, Appl. Phys. Lett. 40, 66 (1982).

(15) Silicon/Metal Silicide Heterostructures Grown by Molecular Beam Epitaxy, J. C. Bean and J. M. Poate, Appl. Phys. Lett. 37, 643 (1980).

(16) Double Heteroepitaxy in the Si(111)/$CoSi_2$/Si Structure, S. Saitoh, H. Ishiwara and S. Furukawa, Appl. Phys. Lett. 37, 203 (1980).

(17) Epitaxial Silicides, R. T. Tung, J. M. Poate, J. C. Bean, J. M. Gibson and D. C. Jacobson, Thin Solid Films 93, 77 (1982).

(18) Schottky-Barrier Formulation at Single-Crystal Metal-Semiconductor Interfaces, R. T. Tung, Phys. Rev. Lett. 52, 461 (1984).

(19) Silicon Molecular Beam Epitaxy on Gallium Phosphide, T. deJong, W. A. S. Douma, J. F. van der Veen, F. W. Saris, Appl. Phys. Lett. 42, 1037 (1983).

(20) Pseudomorphic Growth of Ge_xSi_{1-x} on Silicon by Molecular Beam Epitaxy, J. C. Bean, T. T. Sheng, L. C. Feldman, A. T. Fiory and R. T. Lynch, Appl. Phys. Lett. 44, 102 (1984).

(21) Ge_xSi_{1-x}/Si Strained-Layer Superlattice Grown by Molecular Beam Epitaxy, J. C. Bean, L. C. Feldman, A. T. Fiory, S. Nakahara, and I. K. Robinson, J. Vac. Sci. Technol. A2, 436 (1984).

(22) Fabrication of Metal-Epitaxial-Insulator-Semiconductor-Field-Effect-Transistors using Molecular Beam Epitaxy of CaF_2 on Si, T. P. Smith, J. M. Phillips, W. M. Augustyniak and P. J. Stiles, to be published Appl. Phys. Lett.

(23) Strained-Layer Superlattices from Lattice Mismatched Materials G. C. Osbourne, J. Appl. Phys. 53, 1586 (1982).

(24) Heterostructure Bipolar Transistors and Integrated Circuits, H. Kroemer, Proc. IEEE 70, 13 (1982).

(25) Modulation Doping of Ge_xSi_{1-x} Strained layer Heterostructures, R. People, J. C. Bean, D. V. Lang, H. L. Stormer, A. M. Sergent, K. W. Wecht, R. T. Lynch, and K. Baldwin, submitted to Appl. Phys. Lett.

PROCESS-INDUCED MICRODEFECTS IN VLSI SILICON WAFERS*

Fumio Shimura and Robert A. Craven
Monsanto Electronic Materials Company, St. Louis, Missouri 63137

ABSTRACT

The process by which silicon wafers are transformed into functional VLSI circuitry is complexly related to the starting material properties. This paper will review current understanding of the nature of process-induced microdefects and the use of complementary analytical techniques to elucidate the impact of these defects on circuit yield. Correlations between the occurrence of microdefects and the presence of impurities such as oxygen, carbon, nitrogen, and transition metals will be explored. Techniques for gettering unwanted contaminations away from the active device regions of the silicon wafer will also be discussed.

INTRODUCTION

During the last decade, semiconductor electronics technology has been extensively developed. This development of VLSI technology is based on the fruits of the interrelated research of chemistry, physics, and engineering.

Since the beginning of this year, several Japanese semiconductor device producers have demonstrated production of 1M DRAM and 256K SRAM devices [1] and have also announced that their 256K DRAM production scale has reached 300,000-1,000,000 chips per month[2]. Furthermore, it is expected that the mass production of the 1M DRAM will begin in 1987 [3]. Although the term "VLSI" has been used since 1976 [4], the facts mentioned above suggest that 1984 is really the first year of the VLSI era.

There are many possible causes for VLSI yield loss, but the interrelationship of the device processing with the properties of starting material used for IC fabrication becomes even more important with the onset of the VLSI era. The qualities required for silicon products are very strict to allow achievement of high yield in mass-produced VLSI devices[5,6]. Silicon material modifications which take place during VLSI fabrication are extremely complex. Even in recent high-quality silicon wafers, which are grown without any threading dislocations, various kinds of structural defects are induced during thermal processes [7] causing problems with electrical properties such as junction leakage enhancement, soft breakdown, lifetime degradation, and functional failure due to resistivity shifts [8]. The micron or sub-micron design rules in VLSI devices increase the sensitivity to these microdefects, to say nothing of macroscopic-defects such as slip dislocations and oxidation induced stacking faults from accidental front surface mechanical damage.

* A part of this work was carried out at Fundamental Research Laboratories, NEC Corporation.

This paper reviews process-induced microdefects, which will be classified either as surface or interior microdefects, in silicon wafers subjected to various heat treatments which simulate integrated circuit fabrication processes. These microdefects will be characterized by means of complementary analytical techniques[9]. The emphasis here will be on correlating induced defects with impurities such as transition metals, oxygen, carbon, and nitrogen. Topics discussed include gettering techniques which are widely employed in the semiconductor industry, and ideas for multizoned structures which will enhance the yield and manufacturability of VLSI silicon devices will be proposed.

SURFACE MICRODEFECTS

Contamination with transition metals[10] during thermal processes often results in the induction of microdefects with a density of about $10^6/cm^2$ into the region near the wafer surface. These surface microdefects manifest themselves as small saucer pits (S-pits) after preferential chemical etching. The density of surface microdefects increases drastically with heat treatments at temperatures higher than 1100°C in a steam ambient[11].

Transmission Electron Microscopy (TEM) with energy dispersive X-ray analysis (EDX) has revealed that most of these surface microdefects occur as impurity clusters such as copper, nickel, iron, and chromium[11 14]. However, in heavily contaminated surfaces, small stacking faults ($\leq$ 1µm) associated with precipitates of copper or nickel have been observed by TEM-EDX[15 16]. A diffraction contrast analysis by TEM characterized the stacking faults to be extrinsic-type bounded by 1/3 <111> Frank partial dislocations[15 16].

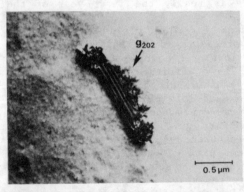

Figure 1. TEM micrograph of stacking fault in a wafer surface after annealing at 1100°C for two hours in steam [15].

Figure 1 shows a surface stacking fault in which a Frank partial dislocation loop is decorated with whisker-like precipitates. Figure 2(a) also shows a TEM micrograph of two different types (arrow-A and arrow-C) of surface-microdefects [9]. The EDX spectra of the defect A and a defect-free region B shown in Figure 2(b) indicates the association of copper with the defect A. On the other hand, Electron Energy Loss Spectroscopy (EELS) of the defect C and the region B is shown in Figure 2(c). The EELS data and the TEM images characterize that the defect C includes both carbon and oxygen, and its morphology is obviously different from that of the defect A. The carbon-oxygen association will be discussed in more detail in the next section.

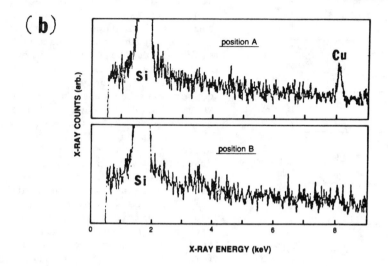

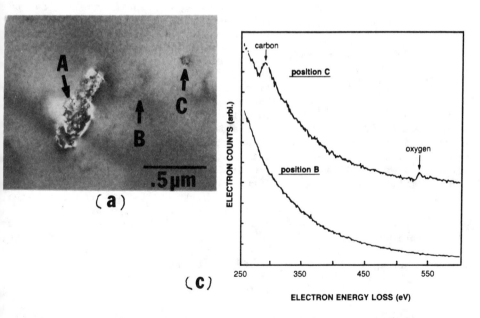

Figure 2(a). TEM micrograph of two types of surface-microdefects, (b) EDX spectra at the defects A and defect-free region B, and (c) EELS spectra of the defect C and the region B [9].

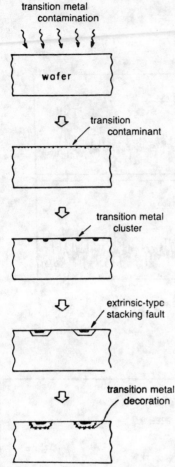

Figure 3. A model for formation and growth of surface-microdefects. Cross-sections of a wafer at stages are schematically illustrated.

The morphology of surface microdefects depends on the degree of contamination and on the nature of contaminants. A recent study[13] has shown that chromium contamination orginates stacking faults, while nickel or iron contamination does not originate stacking faults but only precipitates. And copper may be the major contaminant which associates with these surface microdefects. A schematic model for the formation and growth of surface microdefects is shown in Figure 3. Transition metals adsorbed on a wafer surface agglomerate during thermal processes at a high temperature.
These clusters manifest themselves as S-pits after preferential chemical etching. At this stage, lattice defects such as stacking faults are not formed [11,14]. After further agglomeration of impurities, with the resultant formation of clusters larger than some critical size, extrinsic stacking faults are generated [17,20]. These stacking faults grow by the emission of vacancies or by the absorption of self-interstitials from the surrounding matrix. The stacking fault grows during oxidation since the concentration of vacancies at the oxide-silicon interface is reduced to below the equilibrium value. If contamination with transition metals continue after stacking fault formation, the contaminants will be trapped preferentially at Frank partial loops (see Figure 1) resulting in stabilization of both stacking faults and the contaminants themselves. These secondary precipitates may have higher metallic concentrations than the original clusters. Most EDX work has focused on these latter forms of surface microprecipitates, but many workers are continuing to trace these defects to their early stage.

INTERIOR MICRODEFECTS

Surface microdefects are induced by external contamination, while interior microdefects are primarily caused by precipitation of oxygen. VLSI devices are almost exclusively fabricated on Czochralski (CZ) silicon wafers which may contain oxygen in the order of 10^{18} atoms/cm^3. Regardless of the oxygen concentration, the properties and yield of VLSI devices are extremely affected by the behavior of oxygen in CZ silicon wafers [2].

Since oxygen is usually supersaturated in CZ silicon wafers at modern processing temperatures, heat treatment leads to precipitation of oxygen which results in formation of Si O precipitates. The nucleation and growth mechanism of this oxygen precipitation is not simple, and is still a hot subject of discussion in spite of numerous studies during the last several years [21-25]. Both the growth conditions and the thermal history of an individual crystal must be taken into consideration if a full understanding of this phenomena is to be reached.

In this section, interior microdefects are analyzed with infrared (IR) absorption spectra and TEM images. The effect of subsidiary impurities such as carbon and nitrogen on oxygen precipitation is then discussed.

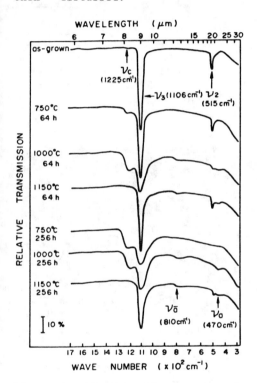

Figure 4. IR absorption spectra for as-grown and heat treated CZ silicon wafers.

The IR absorption spectra ranging from 300 to 1700cm^{-1} for an as-grown specimen and specimens annealed at 750°C, 1000°C, or 1150°C for 64 or 256 hours [26] are shown in Figure 4. For an as-grown silicon specimen, absorption peaks at 515, 1106 and 1225cm^{-1} are observed. The peaks at 515 and 1106cm^{-1} are due to interstitial oxygen [27], while the one at 1225cm^{-1} is due to Si-O precipitates [28,29]. The Si-O absorption at 1225cm^{-1} increases in specimens which have been annealed at 750°C or 1000°C. Furthermore, absorption bands around 470 and 810cm^{-1} which are due to amorphous SiO_2 [30] are observable in specimens annealed at 1000°C or 1150°C. The intensity of Si-O spectra increases inversely with that of interstitial oxygen spectra depending on annealing conditions. The amount of precipitated oxygen depends on many factors [31]. Si-O precipitates begin to dissolve and precipitated oxygen redistributes to interstitial sites by heat treatments at temperatures higher than 1200°C [32-34].

The actual morphology of interior-microdefects depends primarily on the annealing temperatures and heat treatment sequence. Typical TEM images of interior-microdefects observed in CZ silicon wafers subjected to a heat treatment at various temperatures are shown in Figures 5-8. In most cases, low temperature (<750°C) heat treatments generate dense but small precipitates (Figure 5), while high tempera-

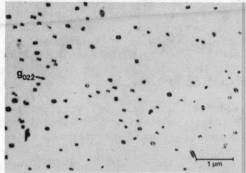
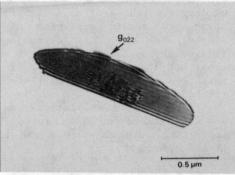

Figure 5. TEM micrograph of Si-O precipitates in a CZ silicon wafer subjected to a 750°C/256 hours heat treatment [33].

Figure 6. TEM micrograph of a stacking fault associated with Si-O precipitates in a CZ silicon wafer subjected to a 950°C/16 hours heat treatment [15].

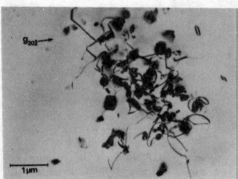

Figure 7. TEM micrograph of precipitate-dislocation-complex (PDC) in a CZ silicon wafer subjected to a 1000°C/256 hours heat treatment [33].

Figure 8. TEM micrograph of an octahedral precipitate of amorphous SiO_2 in a CZ wafer subjected to a 1150°C/64 hour heat treatment [35].

ture (>1100°C) heat treatments generate octahedral precipitates (Figure 8) with low density. Heat treatments at medium temperatures (850-1050°C) generate several kinds of defects (Figures 6 and 7) in a complicated manner. Multistep heat treatments which simulate VLSI processing (i.e. combination of low, medium and high temperature heat treatments) make the interior defect behavior much more complex [33,34]. Oxygen precipitation leads to generation of secondary defects such as dislocations and stacking faults. The type of secondary defects is determined by the density and size of Si-O precipitates after the initial processing steps. Small but highly dense precipitates generate a nearly homogeneous cloud of silicon self-interstitials. This does not tend to form dislocations and stacking faults. When large precipitates, either rectangular platelets or octahedra are formed, however, they tend to emit a very high local concentration of self-interstitials and to nucleate complex precipitate dislocation networks and stacking faults. These complex networks and stacking faults can be extremely effective gettering sites, as discussed in the next section, but can, if carried to an extreme, significantly reduce the strength of the silicon wafer and lead to process induced warpage. A complete theory of the formation of these defect structures has not yet been developed. The complete model will include a microscopic understanding of the role of vacancies, interstitials and secondary impurities in addition to the oxygen precipitate.

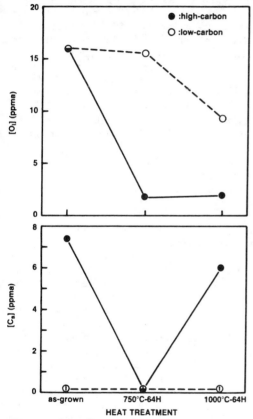

Figure 9. Change in the concentration of interstitial oxygen and substitutional carbon by heat treatments at 750°C and 1000°C.

Carbon, along with oxygen, has been regarded as one of the two major impurities in CZ silicon [36-38] and has been widely observed to enhance oxygen precipitation [39-43]. CZ silicon wafers with a different substitutional carbon concentration $[C_s]$ (<0.05 ppma or 7.3 ppma)* but with the same interstitial oxygen concentration $[O_i]$ (17 ppma)** were subjected to heat treatments at 750°C or 1000°C for 64 hours. The change in $[O_i]$ and $[C_s]$ is shown in Figure 9. The amount of precipitated oxygen in a high-carbon wafer is much larger than in a low-carbon wafer. The effect of carbon is particularly distinctive in a 750°C heat treatment. Furthermore, it should be noted that $[C_s]$ decreased drastically by a heat treatment at 750°C, but not at 1000°C. Carbon atoms became inactive for IR by the heat treatment at 750°C. They are evidently changing site from a substitutional site to a site which is complexed with oxygen atoms or vacancy/interstitial complexes.

*   ASTM   F123-81
**   ASTM   F121-80 (DIN 50435/1)

These results suggest that carbon is involved in the nucleation of oxygen precipitates at 750°C, but not in growth of the precipitates at 1000°C. A microprecipitate shown by arrow C in Figure 2(a) has been characterized by EELS to be associated with carbon and oxygen as shown in Figure 2(c). Although the microprecipitate was observed in a surface region of a different specimen, Figure 2 shows direct evidence of carbon-oxygen association with a defect. These phenomena may be explained by reduction of the interfacial energy of the precipitates either by incorporation of carbon into the precipitate or its agglomeration to the precipitate surface [41].

Based on the previous discussion, there is no doubt that carbon as a subsidiary impurity does play some role in oxygen precipitation. Because of its very small segregation coefficient (0.07 ± 0.01 [41]), carbon is distributed inhomogeneously both axially and radially in a CZ silicon crystal. The effective segregation coefficient is itself strongly dependent on the thermal gradient at the liquid-solid melt interface. Inhomogeneous distribution of carbon may become a problem in VLSI fabrication especially with large diameter (>150mm) wafers.

It should be noted here that the typical carbon concentration in recent electronic-grade CZ silicon wafers is less than 0.05 ppma; which is less than the current detection limit of the IR absorption method. Therefore, whether such a small amount of carbon really affects oxygen precipitation or not is still in question. At the present time, there is no easy way to examine the effect of carbon of less than 0.05 ppma.

This discussion has focused on the role that carbon is playing when it has been grown into the crystal. The effect of carbon or organic contamination on the surface of the silicon wafer is becoming an increasing concern of VLSI device fabricators. Possible repercussions of this contamination may include low reliability of oxides grown on contaminated wafers, or non-uniform oxidation rates for low temperature thin oxide processes.

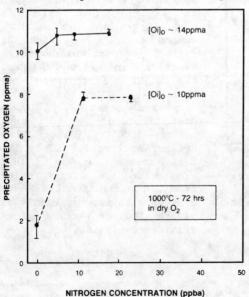

Recently, nitrogen has also been identified as a subsidiary impurity which has both effects on strengthening FZ[45] and CZ[46] silicon wafers, and effects on oxygen precipitation in CZ silicon wafers [46,47]. CZ silicon crystals grown from the melt with and without intentional nitrogen doping have been subjected to various heat treatments. An example which demonstrates

Figure 10. Precipitated oxygen versus the concentration of nitrogen (calculated) in CZ silicon wafers with different initial oxygen concentrations [46].

the nitrogen effect on oxygen precipitation is shown in Figure 10 [46]. The nitrogen enhancement of precipitation is distinct in low-oxygen material in which oxygen precipitation does not normally occur. The effect of nitrogen may be explained by the recent observation of microdefects in a CZ wafer grown from the melt with intentional nitrogen doping [47]. These defects were seen without subsequent thermal processing of the silicon wafer. However, a mechanism which might explain why such a small amount of nitrogen has dramatic effects on silicon properties has not yet been elucidated.

GETTERING

Both the control and reduction of defects in silicon wafer has become more critical as integrated circuit densities increase and larger chips become a reality. Device processing, which is not always perfectly clean from a microscopic point of view, requires the application of gettering techniques to VLSI silicon wafers. Gettering includes two principal effects: one is to suppress process-induced surface microdefects, and the other is to increase minority carrier lifetime in the front surface active device region by capturing or gettering transition metal impurities.

A variety of gettering techniques have been utilized both before, during, and in some cases, after critical integrated circuit process steps [8]. Gettering techniques are classified as two categories: internal (or intrinsic) gettering, and external (or extrinsic) gettering.

Since Tan, et al [48], first reported that interior defects generated during processing getter surface defects, internal gettering techniques[49,50] have been a stimulating subject to study from both the points of VLSI engineering and materials science[22,23]. The principle of internal gettering is that VLSI devices are fabricated in the surface "denuded-zone" of wafers with an internal gettering region beneath this zone which is composed of interior precipitate-induced defects. In addition to two principal effects of gettering mentioned above, internal gettering can increase VLSI parametric yield by minimizing device problems such as memory-function error from alpha-particles or minority carriers caused by impact-ionization, can have positive effects on CCD imager yield by gettering minority carrier generation sites near the front surface, and may be useful in controlling latch-up in conventional CMOS by lowering the recombination lifetime in the substrate adjacent to the complementary SCR transistor.

On the other hand, external gettering uses back-surface damage induced by various techniques such as mechanical abrasion, intense laser beam ablation, ion implantation, phosphorus diffusion and thin film deposition [8]. Regardless of the manner in which the damage is induced, the common intention is to introduce strain into the silicon lattice with the resultant generation of lattice defects which function as gettering sites for impurities during subsequent heat treatments.

To demonstrate these gettering effects, external and internal gettering techniques were applied to fabrication of high-speed bipolar logic device with a shallow emitter-based junction to evaluate the reduction of leakage current between collector and emitter (I_{ce}). These bipolar devices were fabricated using several types of CZ wafers with a reduced-pressure grown epitaxial layer[51]. The resultant yield rates for various wafers used are shown in Table I [21,52]:

TABLE I - Various types of CZ wafers used for high-speed bipolar device fabrication and yield rate ($I_{ce} < 1\mu A$) for each wafer.

Gettering Treatment		Initial Oxygen Concentration (ppma)	Yield Rate (%)	Yield Rate Ratio
MIG		12-15	80.0	6.2
		15-18	70.0	5.9
		18-21	81.8	6.3
EG	MD-Soft*	15-18	26.3	2.0
	MD-Hard**	15-18	38.2	2.9
NONE		15-18	13.0	1

* Soft mechanical abrasion damage
** Hard mechanical abrasion damage

I_{ce} yields in gettered wafers are much higher than ungettered wafers. Comparing the gettering techniques, a multistep internal gettering (MIG) technique [53] is superior to an external gettering technique in the device evaluated. Furthermore, the MIG technique was ascertained to be effective even for wafers with varying initial $[O_i]$. The interior defects and denuded zone observed in the cross-section of MIG-treated wafer are shown in Figure 11.

Back-surface mechanical damage techniques have many disadvantages for application to a reduced-pressure epitaxial growth process for thin layers. In addition to extreme difficulty of controlling particulate generation, the gettering effect is reduced because: i) a part of damaged layer is removed by HCl gas etching prior to epitaxial growth and HCl gas produced from the SiH_2Cl_2 source during epitaxial growth; ii) gas etching on the same layer is accelerated by reduced pressure; and iii) the damaged layer is annealed out during high temperature processes [51]. An MIG technique and/or an enhanced-gettering technique [54] which applies silicon nitride [51] or polysilicon [55,56] backside layer are recommended for VLSI device processing particularly when reduced-pressure epitaxial growth is involved.

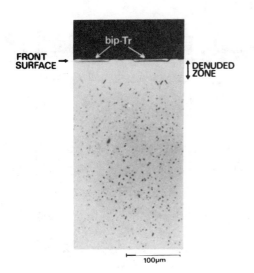

Figure 11. Cross-section of an MIG-treated bipolar transister wafer [21].

Dyson, et al [57], and Borland, et al [58], have emphasized these points in their recent discussions of P/P$^+$ silicon epitaxial layers.

Recent work has focused on the exact mechanism and sites which are responsible for the internal gettering effect. One investigation subjected CZ silicon wafers to a two-step heat treatment. It was found that plastic lattice deformation or dangling sites introduced by dislocations and/or stacking faults are necessary for effective IG sinks. Highly dense Si-O precipitates which cause elastic strain in matrix silicon but not plastic deformation do not result in effective internal gettering [23]. The internal gettering mechanism may be explained by a Cottrell effect [59]. Cottrell has given a theoretical explanation of segregation phenomenon of impurities along dislocations.

There has been a great deal of excitement lately associated with the use of epitaxial substrates in VLSI production. P/P$^+$ epitaxial structures increases the holdtime of DRAM devices [60], and both P/P$^+$ and N/N$^+$ epitaxial structures suppresses latch-up in CMOS circuits[61]. If an IG technique is applied to N$^+$ substrates, certain shortcomings have been expected [62] because of the suppression of oxygen precipitation in heavily doped N-type material via a mechanism involving the coulomb attraction of positively charged dopant and negatively charged silicon self-interstitials [63,64]. Recent studies on gettering techniques for N$^+$ wafers, however, have shown that a multistep internal gettering technique [65] and an enhanced gettering technique which uses a polysilicon backside layer [57] can generate interior defects sufficiently for internal gettering. These results show that internal and external gettering techniques should be combined to optimize the match between individual device properties, the fabrication process, and the silicon substrate.

The recommended multizoned structure consisting of four zones for VLSI silicon wafers is illustrated in Figure 12. VLSI device elements are physically fabricated in the device zone near the wafer front surface. This region must be free of unwanted impurities, structural imperfections and wafer strain. Under the device zone, the denuded zone in which electronic elements function and an internal gettering (IG) zone are formed by the combination of heat treatments including a process of oxygen out-diffusion at a high temperature. To enhance gettering and to resist warpage during thermal cycles, an external gettering (EG) zone with a backside layer of polysilicon, possibly in combination with silicon nitride is recommended.

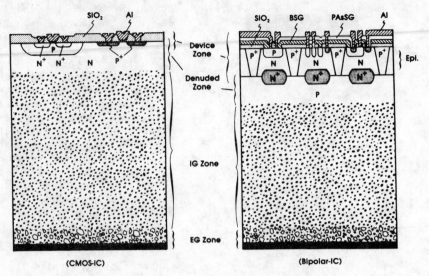

Figure 12. The recommended multizoned structure for VLSI silicon wafers.

SUMMARY

Process-induced microdefects in VLSI silicon wafers are classified as either surface or interior microdefects as summarized in Figure 13. Transition metal contamination during thermal processing initiates surface microdefects. On the other hand, interior microdefects are caused by oxygen precipitation. Si-O precipitates originate secondary lattice defects such as dislocations and stacking faults. Secondary interior defects play a key role as internal gettering sinks for impurities which would otherwise limit device performance or initiate surface microdefects.

Since the silicon modifications which take place during VLSI fabrication are extremely complex, a close cooperation between the silicon crystal producer and the VLSI device producer is required for the successful and profitable development of VLSI technology.

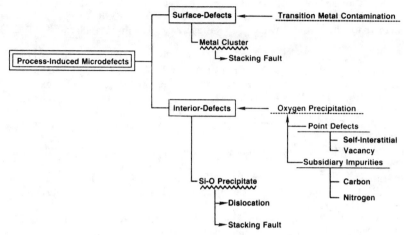

Figure 13. Summary of process-induced microdefects.

ACKNOWLEDGEMENTS

The authors are grateful to all of our colleagues who have contributed to our understanding of VLSI materials. We would like to especially thank Drs. H. Tsuya, P. Fraundorf and H-D Chiou, who have cooperated in the research on which the present review is based.

REFERENCES

1. 1984 IEEE Int'l Solid-state Circuits Conf. (San Francisco, Feb. 22-24, 1984).
2. DENSHI-ZAIRYO, Mar. 1984 (in Japanese).
3. THE YOMIURI SHINBUN, July 3, 1984 (in Japanese).
4. S. M. Sze (ed.), VLSI TECHNOLOGY (McGraw-Hill, New York, 1983).
5. S. Takasu, VLSI SCIENCE AND TECHNOLOGY/1984 (K. E. Bean and G. A. Rozgonyi, eds., The Electrochemical Society, Pennington, PV 84-7, 1984), p. 490.
6. F. Shimura and H. R. Huff, VLSI HANDBOOK (N. G. Einspruch, ed., Academic Press, New York, 1984), to be published.
7. K. V. Ravi, IMPERFECTIONS AND IMPURITIES IN SEMICONDUCTOR SILICON (John Wiley & Sons, New York, 1981).
8. J. R. Monkowski, Solid State Technology/July 1981, p. 44.
9. R. A. Craven, F. Shimura, R. S. Hockett, L. W. Shive, P. B. Fraundorf and G. K. Fraundorf, VLSI SCIENCE AND TECHNOLOGY/1984 (K. E. Bean and G. A. Rozgonyi, eds., The Electrochemical Society, Pennington, PV 84-7, 1984), p. 20.
10. E. R. Weber and N. Wiehl, DEFECTS IN SEMICONDUCTORS II (S. Mahajan and J. W. Corbett, eds., North-Holland, N. Y., 1983), p.19.
11. H. Tsuya and F. Shimura, Phys. Stat. Sol. (a) $\underline{79}$, 199 (1983).
12. F. Shimura, H. Tsuya and T. Kawamura, J. Electrochem. Soc. $\underline{128}$, 1579 (1981).
13. G. K. Fraundorf and R. A. Craven, DEFECTS IN SILICON (W. M. Bullis and L. C. Kimerling, eds., The Electrochemical Society, Pennington, PV 83-9, 1983), p. 406.

14. C. J. Werkhoven, C. W. T. Bulle-Lieuwma, B. J. H. Leunussen and M. P. A. Viegers, J. Electrochem. Soc. 131, 1388 (1984).
15. F. Shimura, H. Tsuya and T. Kawamura, J. Appl. Phys. 51, 269 (1980).
16. W. T. Stacy, D. F. Allison and T-C. Wu, SEMICONDUCTOR SILICON 1981 (H. R. Huff, R. J. Kriegler and Y. Takeishi, eds., The Electrochemical Society, Pennington, PV 81-5, 1981), p. 344.
17. G. R. Booker and W. J. Tunstall, Phil. Mag. 13, 71 (1966).
18. W. K. Tice and T. C. Huang, Appl. Phys. Lett. 24, 157 (1974).
19. S. Mahajan, G. A. Rozgonyi and D. Brasen, Appl. Phys. Lett. 30, 73 (1977).
20. J. R. Patel, K. A. Jackson and H. Reiss, J. Appl. Phys. 48, 5279 (1977).
21. F. Shimura, VLSI SCIENCE AND TECHNOLOGY/1982 (C. J. Dell'Oca and W. M. Bullis, eds., The Electrochemical Society, Pennington, PV 82-7, 1982), p. 17.
22. H. R. Huff, R. J. Kriegler and Y. Takeishi (eds.), SEMICONDUCTOR SILICON 1981 (The Electrochemical Society, Pennington, PV 81-5, 1981).
23. W. M. Bullis and L. C. Kimerling (eds.), DEFECTS IN SILICON (The Electrochemical Society, Pennington, PV 83-9, 1983).
24. J. G. Wilkes, J. Crystal Growth 65, 214 (1983).
25. B. Rogers, R. B. Fair, W. Dyson and G. A. Rozgonyi, VLSI SCIENCE AND TECHNOLOGY/1984 (K. E. Bean and G. A. Rozgonyi, eds, The Electrochemical Society, Pennington, PV 84-1984), p. 74.
26. F. Shimura, Y. Ohnishi and H. Tsuya, Appl. Phys. Lett. 38, 867 (1981).
27. H. J. Hrostowski and R. H. Kaiser, Phys. Rev. 107, 966 (1957).
28. K.Tempelhoff, F.Spiegelberg and R.Gleichmann, in Ref. 22, p.585.
29. S. M. Hu, Appl. Phys. Lett. 36, 561 (1980).
30. E.R.Lippincott, A.van Valkenburg, C.E.Weir and E.N.Bunting, J. Res. Nat. Bur. Stand. 61, 61 (1958).
31. F. Shimura and H. Tsuya, J. Electrochem. Soc. 129, 1062 (1982).
32. F. Shimura, H. Tsuya and T. Kawamura, Appl. Phys. Lett. 37, 483 (1980).
33. F. Shimura, Appl. Phys. Lett. 39, 987 (1981).
34. F. Shimura and H. Tsuya, J. Electrochem. Soc. 129, 2089 (1982).
35. F. Shimura, J. Crystal Growth 54, 589 (1981).
36. H.Föll, V.Gösele and B.O.Kolbesen, SEMICONDUCTOR SILICON 1977 (H. R. Huff and E. Sirtl, eds., The ELectrochemical Society, Pennington, PV 77-2, 1977), p. 565.
37. H. M. Liaw, Semiconductor International/October, 1979, p. 71.
38. P.Stallhofer and D.Huber, Solid State Tech./August, 1983, p.233.
39. U. Gösele and H. Stunk, Appl. Phys. 20, 265 (1979).
40. S. Kishino, Y. Matsushita and M. Kanamori, Appl. Phys. Lett. 35, 213 (1979).
41. R.F.Pinizzotto and H.F.Schaake, DEFECTS IN SEMICONDUCTORS (J. Narayan and T.Y.Tan, eds., North-Holland, N. Y., 1981), p. 387.
42. R. A. Craven, in Ref. 22, p. 254.
43. R.F.Pinizzotto and S.Marks, DEFECTS IN SEMICONDUCTORS II (S.Mahajan and J.W.Corbett, eds., North-Holland, N.Y., 1983), p. 147.
44. T. Nozaki, Y. Yatsurugi and N. Akiyama, J. Electrochem. Soc. 117, 1566 (1970).

45. T. Abe, K. Kikuchi, S. Shirai and S. Muraoka, in Ref. 22, p. 54.
46. H-D.Chiou, J.Moody, R.Sandfort and F.Shimura, VLSI SCIENCE AND TECHNOLOGY/1984 (K.E.Bean and G.A.Rozgonyi, eds., The ELectrochemical Society, Pennington, PV 84-7, 1984), p. 59.
47. F. Shimura, H-D. Chiou, R. Hockett, G. Fraundorf and P. Fraundorf, to be presented at Japan Applied Physics Society Meeting (October, 1984, Okayama).
48. T. Y. Tan, E. E. Gardner and W. K. Tice, Appl. Phys. Lett. 30, 175(1977)
49. R.A.Craven and H.W.Korb, Solid State Technology/July 1981, p.55.
50. D.Huber and J.Reffle, Solid State Technology/August 1983, p.137.
51. K.Tanno, F.Shimura and T.Kawamura, J. Electrochem. Society, 128, 395 (1981).
52. H. Ono, K. Onodera, H. Tsuya and K. Tanno, Proceedings of Japan Applied Physics Society Meeting (1981), 7p-W-7.
53. H. Tsuya, K. Ogawa and F. Shimura, Jap. J. Appl. Phys. 20, L31 (1981)
54. D. C. Gupta, Solid State Technology/August, 1983, p. 149.
55. G. K. Fraundorf, D. E. Hill and R. A. Craven, EMTAS Conf., EE83-131, 1 (1983).
56. W.T.Stacy, M.C.Arst, K.N.Ritz, J.G.DeGroot and M.H.Norcott, DEFECTS IN SEMICONDUCTOR (W. M. Bullis and L. C. Kimerling, eds., The Electrochemical Society, Pennington, PV 83-9, 1983), p. 423.
57. W.Dyson, S.O'Grady, J.A.Rossi, L.G.Hellwig and J.W.Moody, VLSI SCIENCE AND TECHNOLOGY/1984 (K.E.Bean and G.A.Rozgonyi, eds., The Electrochemical Society, Pennington, PV 84-7, 1984), p.107.
58. J.O.Borland, M.Kuo, J.Shibley, B.Roberts, R.Schindler and T.Dalrymple, VLSI SCIENCE AND TECHNOLOGY/1984 (K.E.Bean and G.A. Rozgonyi, eds., The Electrochemical Society, Pennington, PV 84-7, 1984), p. 93.
59. A.H.Cottrell and B.A.Bilby, Proc. Phys. Soc. B62, 49, (1949).
60. R.C.Sun and J.T.Clemens, IEDM Tech. Digest, p. 254 (1977).
61. R. S. Payne, W. N. Grant and W. J. Bertran, IEDM Tech. Digest, p. 248 (1980).
62. C.W.Pearce and G.A.Rozgonyi, VLSI SCIENCE AND TECHNOLOGY/1982 (C.J.Dell'Oca and W.M.Bullis, eds., The Electrochemical Society, Pennington, PV 82-7, 1982), p. 53.
63. A. J. R. deKock and W. M. van de Wijgert, J.Crystal Growth, 49, 718 (1980).
64. A. J. R. deKock and W. M. van de Wijgert, Appl. Phys. Lett. 38, 888 (1981).
65. H.Tsuya, Y.Kondo and M.Kanamori, Jap.J.Appl.Phys. 22, L16 (1983).

TEMPORAL STRUCTURE OF Si (111) SURFACE OXIDATION

K. S. Yi, W. Porod, R. O. Grondin, and D. K. Ferry
Center for Solid State Electronics Research
Arizona State University
Tempe, AZ 85287

ABSTRACT

A temporal Monte Carlo simulation method is used to model the growth on Si (111). The evolution of the oxide is examined as a function of exposure at various temperatues for submonolayer coverages.

Although the oxidation of silicon surfaces has attracted considerable interest,[1-3] the early stages of oxide formation remain controversial.[2-4] In particular, the temporal evolution is not fully understood. Ibach et al.[2] distinguished several different oxygen structures on the surface, through the use of high resolution electron energy loss spectroscopy. In the initial stages of oxidation (at low exposure), these latter authors conclude that three oxygen configurations compete on the silicon (111) surface: (1) bridging atomic oxygen in the silicon-silicon backbond, (2) chemisorbed atomic oxygen, and (3) chemisorbed molecular oxygen.

In recent years, various Monte Carlo (MC) studies of crystal growth[5] and oxide formation[3] have been reported. However, these studies were based upon equilibrium thermodynamics and time entered only as a parameter, the pseudo-time. The pseudo-time in these MC techniques[3] is not the real time measured in the laboratory. The latter is a continuous, well-defined variable that can be related to the above pseudo-time only if realistic kinetic rates are known for the physical processes.

In this paper, we describe the first true dynamic (kinetic) MC model for the evolution of surface oxidation, introducing the time into the model by way of kinetic rates. We start with a two-dimensional ideal clean (111) surface and examine the temporal evolution of the surface layer. Periodic boundary conditions are used for the lateral dimensions. On this ideal and unreconstructed surface, the various processes are simulated for the growth of SiO_x. For example, a deposited oxygen molecule may attach as a chemisorbed state in the form of the peroxy radical. Once attached, it may be evaporated from the surface leaving the site empty, or it can evolve into two chemisorbed atoms through dissociation, or into a chemisorbed atom and a sublayer atomic oxygen (corresponding to the bridging atomic oxygen). By formation of an intermediate ionized state (O_2^-), a chemisorbed molecular oxygen is thought to decay into two atomic species. One of the two atomic oxygens is then incorporated into a Si-Si backbond, forming a Si-O-Si structure below the first layer. This is the MV flip process.[3] The second atom remains as a chemisorbed oxygen in the form of a surface normal bond, or is evaporated. The MV-flip process allows oxygen to penetrate into the sublayer and transforms an oxygen layer into an oxide layer. Each individual process is assumed to occur at a rate which depends upon the detailed local environment of the site. If any of the nearest neighbor or backbond sites are occupied, operations other than O and O_2 evaporations are assumed to be forbidden in the direction of the occupied neighbor. In addition, the influence of the nearest neighbor atoms is included for all atomic processes through mutual coulombic repulsion. Here, atomic oxygen is considered to be ionic, and neighboring ions interact through mutual dipole-dipole repulsion.

The effective kinetic rate of deposition per lattice site is defined by $r_{dep} = \alpha\sigma$, where α and σ are the rate of incidence of oxygen per lattice site and the sticking coefficient, respectively. The evaporation rate is controlled through a thermodynamic identity at equilibrium.[6] Because the absolute values of kinetic rates for surface processes such as migration, dissociation, and the MV-flip process are not available, either from experiments or from a first-principles calculation, they are determined by an appropriate fit to observed growth behavior. The set of rates is adjusted (at a given temperature) so that over-all agreement with experiment is obtained. This fit is also guided by knowledge of relative evaporation rates of O and O_2, an order-of-magnitude estimate of sticking coefficients, and an Arrhenius type temperature variation. The resulting activation energies are substantially lower than those obtained by simple examination of bond-breaking. Hypothetically, this results from the presence of "excess" energy released during preceding processes near a given site. In Table I, the activation energies for the various processes used in this model and the corresponding rates r_i at 700 K are shown.

In the current study, time is introduced by generating a random time for each forthcoming process, which is then stored for each site (e.g. if the ith operation is possible at a given site, a time t_i is assigned for it on that site).[8] A specific process at the site with the smallest t_i is performed and then the next occurrence time and the corresponding operation are updated. Once the local configuration of a site is changed via surface processes, one resets the time of the neighbor sites as well. Results from the current real-time simulations are shown in Fig. 1. Here, we confine ourselves to one monolayer of oxide formation. The number of O-atoms per surface silicon atom of the first layer is plotted as a function of the exposure to molecular oxygen for three different temperatures. Although atomic oxidation is possible, only molecular oxygen is allowed in the initial

Table I. Activation Energies and Transition Rates

Process	E_i (meV)	r_i (700K) (sec^{-1})
O-evap.	370	9.6×10^{-9}
O_2-evap.	215	6.6×10^{-7}
O_2-diss.	140	8.4×10^{-3}
O-migr.	120	9.6×10^{-6}
O_2-migr.	115	1.8×10^{-3}
MV-flip	40	2.9×10^{-2}
deposition	-----	2.9×10^{-1}

gas in this study.[9] The experimental results of Ibach et al.[2,7] are also indicated in Fig. 1. These latter authors suggest that the bridging atomic oxygen (Si-O-Si) is dominant at 700 K for monolayer coverage, i.e. exposure up to 50 Langmuir. The chemisorbed peroxy molecular state is the preferred interpretation of the observed vibrational features at low temperatures (~100K) and a

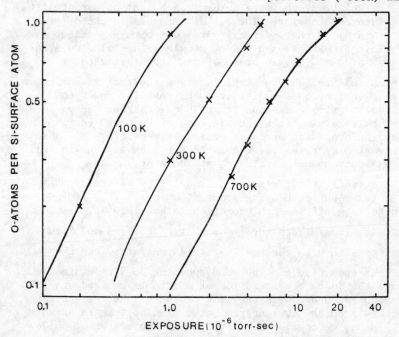

Fig. 1. Ratio of number of oxygen atoms and silicon surface atoms for Si (111) surfaces as a function of oxygen exposure. The crosses are the experimental results of H. Ibach et al.[2,7]

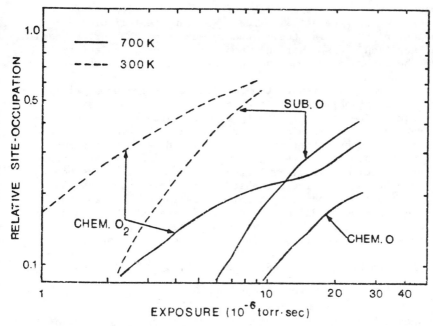

Fig. 2. Relative site-occupation of oxygens in three different local configurations as a function of oxygen exposure. The solid and dashed lines are the results at T = 700K and T = 300K, respectively.

mixed phase of bridging oxygen and peroxy molecule seems to appear at intermediate temperatures.[2,7] Our results are consistent with this interpretation. In Fig. 2, the relative contributions of different oxygen states are shown. These results illustrate that surface oxygen is a mixture of sublayer atoms, chemisorbed atomic, and chemisorbed molecular species. The sublayer oxygen results from the MV-flip. The chemisorbed O_2 participates in the MV-flip and the by-product atomic oxygen is largely evaporated at high temperatures. As the surface coverage approaches unity, the contribution of the peroxy state increases because the sublayer oxygen is expected to attract another O_2 before allowing additional bridging oxygen at the same site. However, sites with more than one sublayer oxygen are observed in the current simulation, which corresponds to the formation of SiO_x units with x>1.[2,4] At 300 K, molecular oxygen is dominant until the surface coverage is unity, which occurs at about 5L. Above ~5L, contributions from sublayer oxygen become considerable and the atomic state remains a minor fraction. At 100 K, chemisorbed oxygen in the peroxy state is the only specie seen in the simulation for coverage up to one monolayer. The formation of an adsorbed oxygen monolayer prior to the oxide formation is only seen at low temperatures. This is in contradiction to the common assumption that oxidation has two distinctive stages: a fast initial adsorption process up to about one monolayer coverage, followed by a slower sorption process to form the actual oxide. We draw the following general conclusions from this study: the evolution of the surface bonding configuration depends strongly upon temperature, and the transition from the adsorbed oxygen layer to the oxide layer starts at a very early stage.

In summary, we report the temporal evolution of Si (111) surface oxidation modeled via a real-time Monte Carlo simulation technique. The temporal structure of the interface oxygen configuration can be explained in terms of kinetic processes with temperature-dependent transition rates. The actual kinetic rates used in the simulation were fit to the experimental data, at a variety of temperatures, by including a simple Arrhenius behavior. It is important to point out that fitting to the data of Ibach et al.[2,7] allows a consistent interpretation of the results, but does not preclude a different choice of rates which could alter the details of the oxidation process. There is a definitive need for measurements and first principles calculations of the kinetic rates themselves.

This work was supported by the Office of Naval Research.

REFERENCES

1. See, e.g., The Physics of SiO_2 and its Interfaces, ed. S. T. Pantelides et al., (Pergamon, New York, 1978) and The Physics of MOS Insulators, ed. G. Lucovsky et al., (Pergamon, New York, 1980).
2. H. Ibach, H. D. Bruchmann, and H. Wagner, Appl. Phys. A 29, 113 (1982), and references therein.
3. S. E. Goodnick, W. Porod, R. O. Grondin, S. M. Goodnick, C. W. Wilmsen, and D. K. Ferry, J. Vac. Sci, Technol. B 1, 767 (1983).
4. See, e.g., G. Hollinger and F. J. Himpsel, J. Vac. Sci. Technol. A 1, 640 (1983), and references therein.
5. J. Singh and A. Madhukar, Phys. Rev. Lett. 51, 794 (1983) and J. Vac. Sci. Technol. B 1, 305 (1983).
6. H. Ibach, W. Erley, and H. Wagner, Surf. Sci. 92, 29 (1980).
7. H. Ibach and J. E. Rowe, Phys. Rev. B 9, 1951 (1974).
8. Here, each event was assumed to occur instantaneously as a Poisson point process with a constant rate function. That is, a time t_i is assigned by t_i = present time $- \ln(\mu/r_i)$ for the ith process, where μ is a random number in $\{0,1\}$ e.g., D. L. Snyder, in Random Point Processes (John Wiley & Sons, New York, 1975).
9. To obtain the results in Fig. 1, the concentration of the initial oxygen molecules was kept constant at $1.4 \times 10^{10}/cm^3$. For the initial sticking coefficient, we used 0.37 at 300 K (with E_i = -15meV), which is slightly larger than the value estimated by J. Onsgaard, W. Heiland, and E. Taglauer, Surf. Sci. 99, 112 (1980).

GETTERING FOR VLSI

G. F. Cerofolini and M. L. Polignano
SGS, 20041 Agrate MI, Milano, Italy

ABSTRACT

The three main gettering techniques (preoxidation gettering on the other side, silicon gettering by segregation, internal gettering) are considered. The techniques are compared in relation to: extended defects, impurities, process compatibility and residual leakage current.

INTRODUCTION

Gettering is a set of fabrication steps able to reduce impurities or extended defects to a negligible amount in active zone of semiconductor devices.
Gettering techniques were already employed in early works; here we quote only a classical work of Goetzberger and Shockley [1] where the technique was set up to reduce soft characteristics of p-n junctions.
At present, gettering techniques are mainly requested in MOS processing because of the large area of MOS devices and the need of high lifetime for dynamic devices, such as dynamic RAMs and CCDs.
Aims of a present-day gettering technique are: 1) the reduction of extended defects in active zones, and 2) the control of lifetime in the wafer.
Though in some situations it could be useful to have a lifetime variable both in depth and along the surface (for instance, in CMOS, to reduce latch-up phenomena, and in dynamic RAMs to make them insensitive to the soft failure by alpha particles), with "control" we mean here an increase of lifetime to values as high as possible.
Three major techniques have been developed to meet the above aims: POGO (pre-oxidation gettering on the other side), SGS (silicon gettering by segregation) and IG (internal gettering).

GETTERING TECHNIQUES

In an MOS process architecture we can imagine that the whole process is formed by three main segments:

1. definition of wells (this may be unnecessary in n-channel devices)

0094-243X/84/1220225-15 $3.00 Copyright 1984 American Institute of Physics

2. definition of insulations and field

3. definition of source and drain.

Any process step at higher level is subjected to heat treatments at lower temperatures than the ones of a process step at lower level.
If the gettering technique is superimposed to an MOS process and one is not allowed to modify dopant profiles, then it must logically be placed either at the zeroth level, where it must sustain even strong heat treatments, or at the fourth level, where however must require only weak heat treatments.
The POGO technique is a gettering technique of the zeroth level[2], while SGS is a gettering technique of the fourth level[3]; IG is a gettering technique operating starting at the zeroth level and completing at the fourth level[4].

POGO

POGO is based on the thermodynamic property of large defects which tend to enlarge at detriment of small defects. Thus the formation of a heavily damaged backside (e.g., by sand blasting, LASER melting, polysilicon deposition) furnishes the preferential sites for gettering embryos of extended defects in active regions.
Fig. 1, A to C, shows the scanning electron microscope (SEM) images of slice backside after cleavage and Secco etching. Fig. 2 shows the back before oxidation.
Fig. 1A and B show the images of the same wafer before and after an MOS process, respectively; the strongest heat treatment of the process is a steam oxidation at 920°C for 6.5 h. Fig. 1B shows that a crown of dislocations is formed at the back and is confined within a layer about 10 μm thick. For another kind of back damage, dislocations are not confined close to the back and are much less dense (fig. 1C). Since both dislocations and grain boundaries are effective getter sites for heavy metals, the POGO endows the back with a number of getter sites.
Backside damage can also be obtained with other techniques; e.g. by phosphorus doping at high temperature (say 1050°C) and low sheet resistivity ($\simeq 5 \; \Omega/\square$), and by depositing a $Si_3 N_4$ layer on the back and hence heating at a temperature so high that the stress is absorbed by silicon by plastic deformations. In both cases the back is endowed with a dislocation crown and this is reached by strong heat treatments. Because of this, POGO can be seen as a zeroth level process step.

227

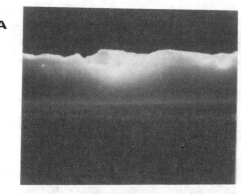

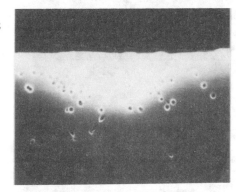

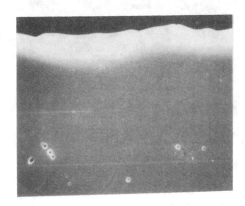

———— 8 μm

Fig. 1. SEM images of wafer backsides after cleavage and Secco etching. Fig. 1A shows the back before processing; fig. 1B shows the back of the same wafer after a prolonged steam oxidation (920°C, 6.5 h) -- a crown of dislocations surrounds each trace of mechanical damage; fig. 1C shows the back of a wafer with different mechanical damage after the same oxidation -- dislocations are less dense and not confined close to the back.

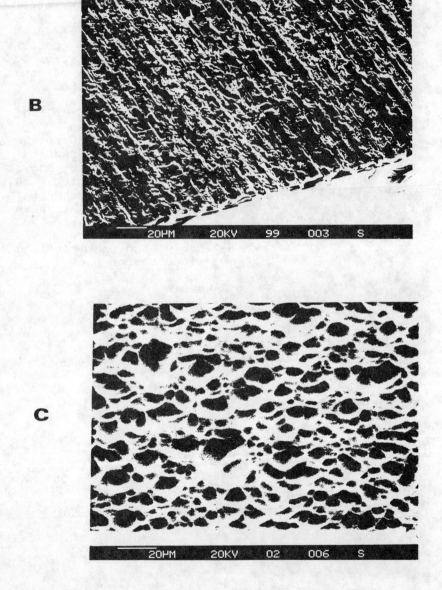

Fig. 2. SEM images of the backs of wafers B (brushed) and C (sand blasted) of fig. 1, before any oxidation step.

SGS

SGS is based on the existence of a segregation factor between a heavily doped (or strongly damaged) region and bulk silicon. Getter sites (which may be phosphorus or boron atoms, or extended defects such as dislocations or oxygen precipitates) must be confined out of active regions, for instance in the contacts, in scribe lines or at the back.

For gold in the Si:P and Si:B systems, the segregation factor between doped and undoped regions is higher the lower the temperature. Heat treatments of the order of $800°C$, 1 h or $700°C$, 4 h are weak enough not to produce dopant profile variation but long enough to complete segregation.

Lower temperatures, say $450°C$, could require times too long, for which other phenomena (thermal donor activation) appear [5].

Because of its gentle heat treatment, SGS can be regarded as a fourth level process step, necessarily preceding heat treatments around $400°C$ characteristic of alloying and dielectric deposition.

IG

Somewhat intermediate is the IG technique. This technique was developed mainly to control oxygen in Czochralsky grown silicon cristal. Oxygen is the major impurity of silicon and its concentration exceeds solid solubility even at temperatures as high as $1200°C$. Oxygen content in active regions (width $\approx 10\ \mu m$) can be reduced by evaporation from the surface and precipitation in the bulk. This process produces therefore an oxygen free layer (the "denuded zone" DZ) close to the surface and a highly damaged region in the bulk. The thickness of the DZ is of the order of the oxygen diffusion length associated with the heat treatment, and size and distribution of oxygen precipitates depends upon various factors such as preexisting nucleation centers, oxygen concentration, temperature and time.

In practice, though the theory of the formation of the DZ is rather simple, several unknown factors make such a process difficult. Two major variants of IG are known -- the LO-HI technique (first step at low temperature, $750°C$, second at high temperature, say $1100°C$) suitable for low oxygen content, and the HI-LO technique (same heat treatments in the opposite order) suitable for high oxygen content. In spite of this, IG is still far from being a reliable technique. For instance Murray [6] poses three questions: 1) what effect does the order of the heat treatments have on denuded zone formation? 2) What effect does a thin oxide layer have on the denuded zone?

3) How reproducible is the denuded zone? After his experimental study, Murray reaches the conclusions that: (a) the results will vary between different vendors, and (b) wafers with the same initial oxygen concentration supplied by different vendors will precipitate differently.

Oxygen precipitation phenomena may be associated with injection of vacancies and interstitials in silicon.

Fig. 3 shows the SEM image of a high oxygen concentration Czochralski crystal (for which the recipe for the DZ formation is a HI-LO process) after a LO-HI process ($750^{\circ}C$, 4 h, N_2, plus $1050^{\circ}C$, 8 h, dry O_2), cleavage and Secco etching. The resulting etch pattern shows a giant bulk defectiveness, due to: oxygen precipitates (the structure of which cannot be resolved by SEM), estrinsic stacking faults (ESF, due to precipitation of self-interstitials), and (?) intrinsic stacking faults (ISF, due to precipitation of vacancies). A different evidence for ISF associated with oxygen precipitation has been given by Claeys et al.[7]

Once the DZ and precipitates are formed, we have a non defective surface zone and a highly defective bulk zone, that allows heavy metals to be segregated far from active zones by anneal at moderate temperature (see above).

CAUSES FOR LEAKAGE

Extended defects (dislocations, stacking faults,...) decorated with heavy metals are usually responsible of soft characteristics, in turn associated with loss of yield.

Extended defects are usually avoided by starting with non-defective materials and operating at moderate temperature with thermal ramps. The POGO may be useful to reduce the concentration of extended defects in active zones.

Once the extended defects are not present, the residual major causes responsible for leakage current are:

1. generation-recombination at the contacts

2. generation-recombination centers

3. donor-acceptor twins

The generation current due to interface states at the $Si-SiO_2$ interface is negligible in most cases.

Contacts can be regarded as elements with infinite generation-recombination rate, so that the minority carrier concentration is therein always the equilibrium one. A contact can be ignored in the description of a device provided that its distance x_c from the depletion

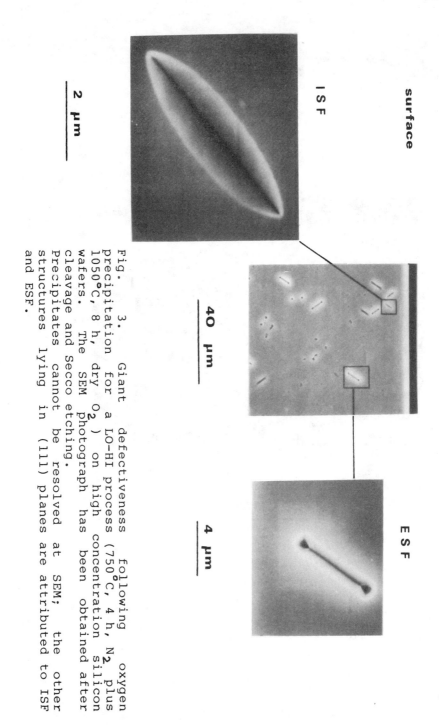

Fig. 3. Giant defectiveness following oxygen precipitation for a LO-HI process (750°C, 4 h, N$_2$ plus 1050°C, 8 h, dry O$_2$) on high concentration silicon wafers. The SEM photograph has been obtained after cleavage and Secco etching. Precipitates cannot be resolved at SEM; the other structures lying in (111) planes are attributed to ISF and ESF.

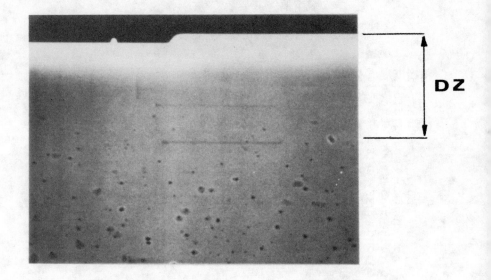

Fig. 4. Formation of the DZ after a LO-HI process (750°C, 4 h, N_2 plus 1050°C, 2 h, dry O_2) on low concentration silicon wafers.

layer is much higher than the minority diffusion length L, $x_c/L \gg 1$.
Since the diffusion length is affected by the concentration of generation-recombination centers, for a given geometry the above inequality can be satisfied or not according to the device process.
In an experimental study of IG, we have compared diodes gettered by SGS technique only with diodes gettered by a LO-HI IG plus SGS technique; diodes of the latter type were found to have a DZ with width around 10 μm (see fig. 4).
The concentration of generation-recombination centers in the denuded zone of the LO-HI diode was found roughly equal to that of the SGS diode; the diffusion length for such a concentration (L = 3800 μm) being higher than the wafer thickness (525 μm). The diffusion current density J_{dif} flowing in the diodes differed from one another by one order of magnitude. Writing J_{dif} in the usual form,

$$J_{dif} = \frac{e D n_i^2}{c_b \ell}$$

where e is the unitary positive charge, D the minority diffusion coefficient, n_i the intrinsic carrier concentration, c_b the substrate dopant concentration, the length ℓ responsible for the diffusion current in the SGS diode is of the order of the wafer thickness, while in the LO-HI diode the length is of the order of the width of the DZ. This fact suggests that the precipitates can be regarded as centers with very high generation-recombination rate. It follows that the minimum width of the DZ is dictated by the specification on the leakage current.

The original Shockley theory of the p-n junction relates the current density J to the applied voltage V through the relationship

$$J = J_{dif} \left[\exp\left(\frac{eV}{k_B T}\right) - 1 \right] \qquad (1)$$

where k_B is the Boltzmann constant and T is the absolute temperature. In absence of a good gettering technique eq. [1] describes accurately only germanium junctions but not silicon junctions. The reason for this drawback is the existance of amphoteric impurities which behave as generation-recombination centers. The generation-recombination rate at these centers, described by a theory due to Shockley, Read and Hall (SRH) [9,10], modifies the $J - V$ characteristic to the form

$$J \propto (|V| + \Phi_{bi})^{1/2} \quad \text{for } |V| > k_B T/e \text{ (reverse bias)}$$

and

$$J \propto \exp(eV/m k_B T) \quad \text{for } V > k_B T/e \text{ (forward bias)},$$

Φ_{bi} being the built-in potential and m an ideality coefficient, which should be around 2, $m \simeq 2$.
Fig. 5 shows the $J - V$ characteristics at 300 K temperature of three diodes gettered with the SGS technique and final anneal at 650°C. The figure clearly shows that all diodes are ideal for forward bias (ideality coefficient m = 1.01+0.01) while for reverse bias only diode A can be considered ideal, while diodes B and C behave approximately as $J \propto |V|$. Diodes of the latter type will be referred to as almost ideal.
Typical lifetimes obtained from reverse characteristics and which result after a good gettering technique are of the order of $3 \cdot 10^{-2}$ s, associated with diffusion lengths of the order of $10^3 - 10^4$ μm, i.e. higher than wafer thickness. This ensures the scarce relevance of SRH generation recombination centers on the $J - V$ characteristics of silicon p-n junctions. About the nature of residual SRH centers, our opinion is that they involve oxygen, which is known to give an SRH center with trap energy of 160 meV from conduction band.

Almost all diodes processed with the SGS technique are ideal for forward bias; their reverse characteristics, however, have a seemingly Ohmic contribution to the current in addition to [12] J_{dif}.
The seemingly Ohmic contribution is of the type

$$I - I_o \propto (|V| + \Phi_{bi})/R$$

where R is the apparent resistance and I_o is the current extrapolated at $V = 0$.
The apparent resistance R has the following features:

1. irrespective of the final anneal temperature

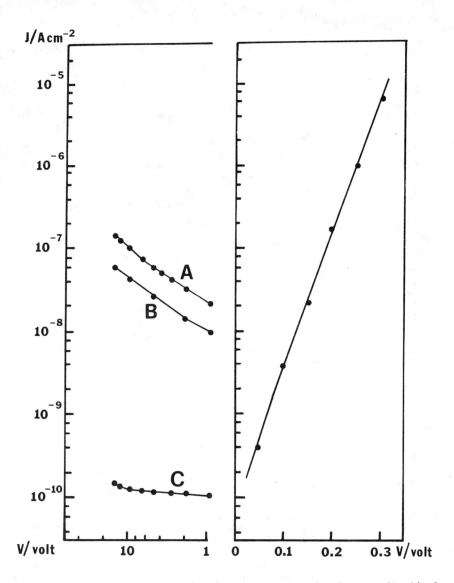

Fig. 5. Experimental $J - V$ characteristics of diodes processed with the SGS technique. Characteristic A is ideal both for forward and reverse bias; characteristics B and C coincide with A for forward bias, but for reverse bias they have a seemingly Ohmic contribution.

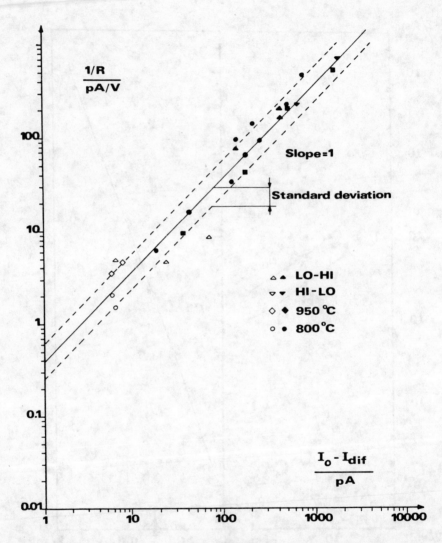

Fig. 6. Log-log plot of $1/R$ vs. $I_o - I_{dif}$ where I_o is the current extrapolated at 0 V. With empty symbols we have denoted diodes which are representative of the average behavior; while with full symbols we have denoted diodes which belong to the tail of the distribution in leakage current.

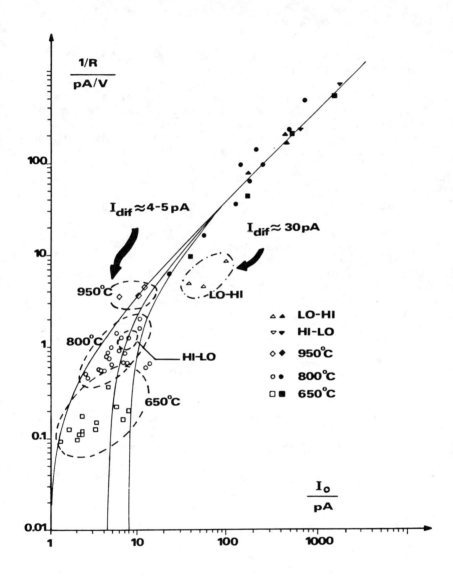

Fig. 7. Log-log plot of $1/R$ vs. I_o. Symbols have the same meaning as in fig. 6.

$$1/R \propto I_o$$

over about four orders of magnitude of I_o; this fact is confirmed by fig. 6;

2. most diodes with final anneal at the same temperature T_{ann} have the same apparent resistance so that a function $R(T_{ann})$ can experimentally be found; fig. 7 shows the values of R and their dispersion at various anneal temperature.

These features are accurately explained by admitting that [15] the extracurrent is due to pure generation (not recombination) centers provided that they are:
i. shallow levels
ii. globally neutral (i.e., donor-acceptor twins)
iii. field assisted and thermally activated
iv. equilibrium defects

The ionization process of a neutral DAT is schematically shown in fig. 8. This scheme suffices to explain the results shown in figs. 5 to 7, though the nature of such centers has not yet been clarified.

Fig. 8. The proposed process for DAT ionization and neutralization.

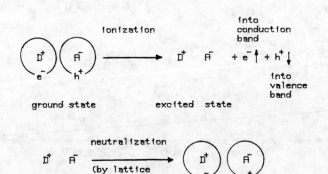

CONCLUSIONS

The principal gettering techniques are discussed in relation to their action mechanism and to their compatibility with an MOS process.
The residual causes of leakage current (SRH centers, contacts and donor-acceptor twins) are considered in relation to a gettering procedure.

REFERENCES

1. A. Goetzberger and W. Shockley, J. Appl. Phys. 31, 1821 (1960)
2. G. A. Rozgonyi, P. M. Petroff and M. H. Read, J. Electrochem. Soc. 122, 1725 (1975)
3. L. Baldi, G. F. Cerofolini, G. Ferla and G. Frigerio, Phys. Stat. Sol. A 48, 523 (1978)
4. T. Y. Tan, E. E. Gardner and W. K. Tice, Appl. Phys. Lett. 30, 175 (1977)
5. V. Cazcarra and P. Zunino, J. Appl. Phys. 51, 4206 (1980)
6. E. M. Murray, J. Appl. Phys. 55, 536 (1984)
7. C. Claeys, H. Bender, G. Declerck, J. Van Landuyt, R. Van Overstraeten and S. Amelinckx, in: Aggregation Phenomena of Point Defects in Silicon (E. Sirtl and J. Gorissen, eds.), The Electrochem. Soc., Pennington, NJ (1984) p. 74
8. W. Shockley, Bell System Tech. J. 28, 435 (1949)
9. W. Shockley and W. T. Read, Phys. Rev. 87, 835 (1952)
10. R. N. Hall, Phys. Rev. 87, 387 (1952)
11. C. T. Sah, R. N. Noyce and W. Shockley, Proc. IRE 45, 1228 (1957)
12. G. F. Cerofolini and M. L. Polignano, J. Appl. Phys. 55, 579 (1984)
13. G. F. Cerofolini and M. L. Polignano, J. Appl. Phys. 55, 3823 (1984)

TWO-DIMENSIONAL NUMERICAL SIMULATION OF THE CHARGE
TRANSFER IN GaAs CHARGE-COUPLED DEVICES OPERATED
IN THE GHz RANGE*

D. SODINI, A. TOUBOUL, D. RIGAUD
Centre d'Electronique de Montpellier
(associé au C.N.R.S., UA 391)
Université des Sciences et Techniques du Languedoc
34060 Montpellier Cedex, France.

ABSTRACT

In a feasability study on bulk channel GaAs C.C.D. used in the GHz range, we show a numerical simulation of the dynamic behaviour of these devices, allowing us to find transfer inefficiency.

The optimal signal charge to be transfered is first calculated taking into account the technological and electrical parameters which characterize the structure to ensure at the same time, a bulk storage and transfer.

The influence of the technology (3 and 4 phases devices, epitaxial or ion-implanted channel, Schottky or MIS gate) and of electric parameters (clock frequency, inter electrode gaps, oxide charges) on the transfer inefficiency value is studied in detail.

INTRODUCTION

The operation of a bulk channel GaAs C.C.D. can be described on the basis of a numerical analysis. This approach involves the numerical solution of the Poisson/continuity equations. A finite difference method has been choosen. As these devices are designed to work in the GHz frequency range, the clock signals have sinusoïdal wave forms. The most significant parameter obtained from these calculations is the transfer inefficiency ε. Furthermore we shall present results related to C.C.D.'s of different technologies such as Schottky gates, MIS gates, epitaxial and ion-implanted channels.

DESCRIPTION OF THE DEVICE

A schematic view of the C.C.D. is given in Figure 1. The storage and the transfer of the signal carriers take place in the n-type channel which is assumed to be fully depleted. An elementary cell comprises 3 or 4 gates (according to the number of tranfer phases) separated by inter-electrodes gaps. According to the kind of technology used, the channel can be controlled by Schottky gates or by a layer of anodic GaAs oxide where metal gates are located. The n-type channel can be created either by epitaxy or by the implantation of Si in a semi-insulating GaAs substrate (which will be treated

* Work supported by C.N.E.T. (Contrat n° 82-6B043-PAB) and the
  C.N.R.S. (Greco Micro-Ondes).

as an equivalent p-type substrate).

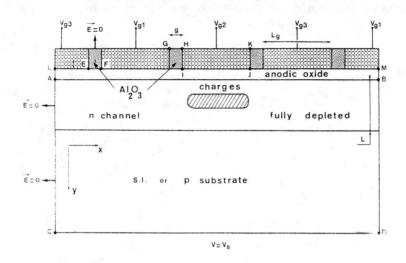

Fig. 1. Schematic cross-section of a B.C.C.D. with boundary conditions used in the numerical solution of the Poisson-continuity equations.

In Figure 2, we have reported the different gate biases for a 3 and a 4-phase B.C.C.D.; the clock signals vary from 0 to 5 volts.

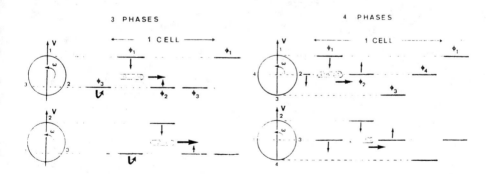

Fig. 2. Transfer mechanism for 3 and 4 phases B.C.C.D.

From now on one can see that the 3-phase device will ensure satisfactory storage conditions due to the higher values of the gate biases when compared to a 4-phase system. Whereas for this 4-phase system, the charge motion should be improved by the variation of the two adjacent gates in the transfer direction.

CONSIDERATIONS ON THE STORAGE OF AN OPTIMAL SIGNAL CHARGE

When using an elementary cell of two half-gates separated by an interelectrode gap (see Fig. 3.a), the optimal signal-charge must be such that the three following conditions are satisfied:

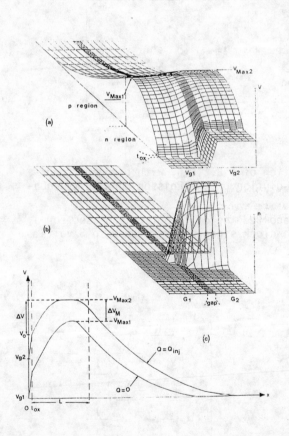

i) The stored charges under the biased G_2 gate must not overflow under the gate G_1 which is not biased. This situation has been shown Fig. 3.b.: it shows clearly that the "well" under G_2 must be deeper than the one under G_1 such that the difference ΔV_{max} defined as $V_{max\,2} - V_{max\,1}$ (see Fig. 3.a and 3.c) must be kept positive.

ii) All contact of the carriers with the interface traps and any reduction of the mobility due to the crystalline defects of the interface must be avoided. So $V_{max\,2}$ will always be greater than the surface potential V_0 under the G_2 gate.

iii) The electric field value at the interface must be less than the breakdown value of the gate oxide and less than the impact ionization field of the GaAs.

Fig. 3. Elementary cell used for the study of the charge handling capacity
a) potentials; b) charge location;
c) 1-D potentials under the gates G_1 and G_2.

The numerical solution of the 1-D Poisson equation and these three conditions allow us to define curves which define for a set of technological parameters the regions of optimal behaviour of the C.C.D.[1,2]

It can be shown that the storage capacity is all the more important when the doping level or the dose and when the channel thickness or the range are reduced. The Schottky gate devices ensure a higher storage than the M.I.S. gate devices under the same operating conditions but require deeper implantations. Notably, a 2-dimensional analysis shows that an implantation at 0.2 μm brings about the appearance of charges on the surface.

STUDY OF THE TRANSFER : NUMERICAL SOLUTION OF THE POISSON/CONTINUITY EQUATIONS

The uncoupled solution of these two equations has been done with the aid of a finite-difference method and a variable mesh size grid. The evolution of the signal charge can be obtained at any time and at any point of the channel.

The physical model of the C.C.D. can be described on the basis of the following assumptions :
i) the n-channel is fully depleted
ii) the carriers of the p-type substrate are such that

$$p(x) = N_A \exp(-\beta V(x)) \quad \text{with } \beta = q/kT \qquad (1)$$

iii) trapping and G-R mechanisms are neglected
iv) one type of carrier is taken into account
v) the displacement current has been calculated : it can be neglected
vi) no hot carriers are found in the signal charge; in equilibrium, the Einstein's relation remains valid and the mobility is but a function of the channel doping level.

The continuity equation solution needs to solve the Poisson equation which can be written as :

$$\nabla^2 \psi = -q/\varepsilon (N_D - N_A - n + p) \qquad (2)$$

The current equation is given by :

$$\vec{J}_n = q (\mu_n n \vec{E} + D_n \vec{\nabla} n) \qquad (3)$$

Using the quasi Fermi level of the signal carriers ϕ_n, one can write :

$$n = n_i \exp \beta (\psi - \phi_n) \qquad (4)$$

and the current expression becomes : $\vec{J}_n = -q n \mu_n \vec{\nabla} \phi_n$ (5)

The continuity equation $q \, \partial n/\partial t = \nabla \cdot \vec{J}_n$ has been expressed, when using the equation (5) and the variables ϕ_n, ln n, as :

$$\frac{\partial}{\partial t}(\ln n) = -(\mu_n \vec{\nabla} \ln n \cdot \vec{\nabla} \phi_n + \mu_n \nabla^2 \phi_n) \qquad (6)$$

If n is expressed as a function of ψ and ϕ_n (eq. 4), the continuity equation becomes :

$$\frac{\partial}{\partial t}(\psi - \phi_n) = -(\mu_n \overrightarrow{\nabla(\psi-\phi_n)} \cdot \overrightarrow{\nabla\phi_n} + \frac{1}{\beta}\mu_n \nabla^2 \phi_n) \qquad (7)$$

In these last equations, we use ϕ_n or ln n rather than n(x) which can vary up to several orders of magnitude, (particularly at the edges of the charge packet) whereas the variations of ϕ_n or ln n are more gradual and improve the stability of the numerical methods used to solve the continuity equation.

For that purpose we have used an iterative scheme (linearized Crank-Nicholson method). The stability of its solution requires a time step as small as 2.10^{-14} second.

TRANSFER RESULTS

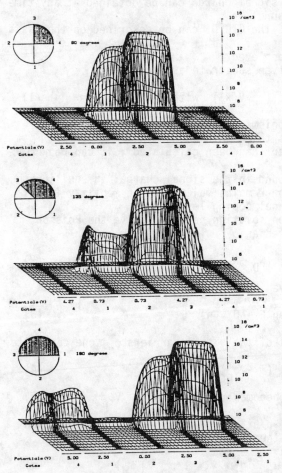

An example of the results of the transfer versus time is given (Fig. 4) for a Schottky technology on an epitaxial n-type channel ($N_D = 2.10^{16}$ at/cm^3). A signal charge of 3×10^{-10} C/m is located at t=0 under the gate G_2 of an elementary four gates-cell. The gate length is 5 μm and they are separated by a 0.5 μm gap; the thickness of the channel is 2 μm. We give the charge location at different steps of the transfer process when the clock frequency is 2 GHz ($T_o = 5\ 10^{-10}$ s). At the time $T_o/4$ the gate number 3 becomes the storage gate (Fig. 4.a) and plays the same role as gate number 2 at t=0. We can see that the charge transfer is incomplete and that about 2 % (logarithmic scale) of the packet is still under gate number 2. The Fig. 4.b,c show the cleavage tendancy of the charges into two parts : the majority of the

Fig. 4. Charge stransfer vs time at 2 GHz for an epitaxial channel 4 phase C.C.D. with Schottky gates.

carriers shift in the transfer direction while the other part on the left hand side flows back to form the transfer inefficiency. This part concerns about $4 \cdot 10^{-6}$ of the initial charge packet. The stored charge under the gates number 3 and 4 on Fig. 4.c is the same as the one which was under the gates number 2 and 3 (Fig. 4.a).

The results concerning a 3-phase device with the same technological parameters are plotted in Fig. 5 for the same clock frequency.

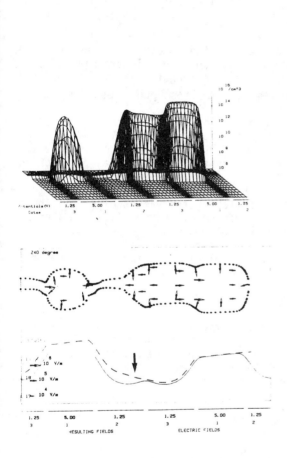

At $t = 2 T_0/3$ (240°), the charge packet is divided into two parts : the main part being located under G_3 and G_1 and a trailing packet under G_1 at the left of the emitting gate G_2 which flows backward. The value of the transfer inefficiency is less satisfactory than for a 4-phase device : ε is equal to $3 \cdot 10^{-4}$. In the same figure, we have plotted the location of the charge packet in the bulk of the n-type channel.
The horizontal arrows give the direction and the amplitude of the resulting field defined as $\overrightarrow{\nabla \phi_n}$ and the others are related to the electrostatic field which ensures the cohesion of the charge packet.
The variation of the quasi Fermi level ϕ_n along the transfer direction is reported in the same figure (dotted line) whereas the electrostatic potential ψ is represented by the other curve : it should be noted that under the gate G_1, where the charge packet has been transferred, the quasi Fermi level tends to be horizontal (thermal equilibrium). Under the other gate G_1, where the

Fig. 5. Charge location for a 3 phase C.C.D. at $t = T_0/2$ (2 GHz) and electric and resulting fields.

residual charges are located, the quasi Fermi level tends to be also horizontal but with an other value. The fields $\overrightarrow{\nabla \psi}$ et $\overrightarrow{\nabla \phi_n}$ are different when a strong carrier concentration gradient exists involving an important diffusion term (see the arrow on Fig. 5.c).

For a 4 phase device the influence of the clock frequency on the transfer inefficiency has been pointed out; the values of ε for different frequencies are summarized in the following table :

Table I Transfer inefficiency versus frequency

f(GHz)	2	3	4
ε	4.10^{-6}	$2.4\ 10^{-4}$	$2.3\ 10^{-3}$

These data must be associated to the incomplete transfer of the free carriers and do not take into account any bulk trapping effects which would lead to higher values of the transfer inefficiency.

These values are in agreement with those published by Deyhimy[3,4] for similar devices.

The same study has been made on ion-implanted channel C.C.D. for different implant range ($0.5\,\mu m - 0.2\,\mu m$) and dose ($1\ 10^{12} - 1.5^{12}$ at/cm^2). At 2 GHz the value of the transfer inefficiency is more important (about $5\ 10^{-2}$ for a 4 phase system) in all cases than for an epitaxial channel.

For epitaxial and ion-implanted channels using a MIS gate technology, we find quite similar results for the values of ε to those obtained by Schottky gates. The presence of charges ($10^{12}/cm^2$) on the oxide-semiconductor interface does not modify these results.

The interelectrode gap value in the range of $0.15 - 0.5\ \mu m$ does not appear as an important technological parameter for the transfer inefficiency.

CONCLUDING REMARKS

Independently of the full compatibility problem for input/output stages the best transfer efficiency is obtained with an epitaxial channel and it seems that the choice of the gate system (MIS or Schottky) is not important. On the contrary, for the best VLSI integration an ion-implanted channel is preferable because of its better storage capability.

REFERENCES

1. K. Torbati, thesis, Montpellier, (1984).
2. D. Rigaud, D. Sodini, A. Touboul, European GaAs I.C. Workshop, (1984), Nice, France.
3. I. Deyhimy et al., Appl. Phys. Lett. 36, 151 (1980).
4. R.C. Eden, I. Deyhimy, Optical engineering, 20, 947, (1981).

MODELLING THE INITIAL REGIME
OF DRY THERMAL OXIDATION OF SILICON

G. Ghibaudo, A. Fargeix
Lab. PCS, ERA-CNRS 659, ENSERG,
23 Av Des Martyrs 38031 Grenoble FRANCE

ABSTRACT

This communication deals with the initial regime of dry thermal oxidation of silicon occurring in the critical thickness range 25 - 50 nm. It is mainly emphasized that the decrease of the diffusivity of the oxidizing species near the Si-SiO$_2$ interface due to high compressive stress is responsible for the anomalously high oxidation rate and for the change with oxidation temperature of the effective activation energy of the parabolic kinetic constant. In addition, the stress levels derived from the kinetic data analysis are found in agreement with those directly determined by Si wafer curvature measurements or deduced from refractive index assessments.

INTRODUCTION

Although the well-known Deal-Grove model of oxidation is satisfactory for steam oxidation, it is quite unsuitable for the early regime of some hundreds of angstroms of dry Si oxidation relevant in the scope of VLSI oxide thickness range. As previously discussed,[1] this fast initial regime has to be attributed not to an enhanced diffusivity near the Si-SiO$_2$ interface, as generally advanced, but to a decrease of the diffusivity in the vicinity of the interface due to high compressive stress within the oxide.

As established before,[2] a new model of oxidation which takes account of a stress dependent diffusivity of the oxidizing species and of a Maxwellian stress relaxation by viscous flow enables the modelling of kinetic oxidation rates as a function of oxide thickness over a wide range of temperatures (700 to 1000°C). It also predicts the different stress levels in the oxide layer and related quantities such as Si wafer curvature or refractive index during and after growth.[3]

Furthermore, as shown recently, the temperature dependence of the parabolic kinetic constant K_p is also well accounted for by the model throughout the free energy change via the modification of the activation energy of the oxidizing species' diffusivity.[4]

In this paper we give an overview of the different features derived from the stress-state model of oxidation and concerning the silicon oxidation kinetics and the stress levels within thermally grown SiO$_2$ films.

FORMULATION OF THE MODEL

In this section we briefly summarize the basic features of the stress state model of oxidation. The main physical processes involved in this model are as follows:[2]

(i) a Deal-Grove process i.e. diffusion of the oxidizing species before reacting at the Si-SiO$_2$ interface,
(ii) an inhibition of the oxidizing species diffusivity due to high compressive stress,
(iii) a stress relief mechnaism via viscous flow.

Thus the inverse growth rate $r = dt/dX_{ox}$ is obtained by integration of the equation:[3]

$$\frac{dr}{dX_{ox}}(X_{ox}) = \frac{1}{vCD[\sigma(X_{ox})]} - \frac{1}{vC}\int_0^{X_{ox}} \frac{1}{D^2} \frac{dD}{dX_{ox}}(x,X_{ox})dx \quad (1)$$

with the initial rate being $r(0) = 1/K_1$ (K_1 is the linear kinetic constant), and where v is the volume change due to oxidation, C is the outer concentration of oxygen and $D[\sigma(x,X_{ox})]$ is the local effective diffusivity of the oxidizing species depending on the local stress level $\sigma(x,X_{ox})$ in the oxide layer of thickness X_{ox} by the relation:[4]

$$D = D_0 \exp[-\frac{\sigma \Delta V}{kT}] \quad (2)$$

where ΔV is the diffusion volume change due to stress and D_0 is the diffusivity of the considered species in the unstressed oxide. In first analysis ΔV can be derived from a compressibility relationship.[4]

The parabolic kinetic constant K_p, roughly equal to $2vCD$, is thus given by:[4]

$$K_p(e_{ox}) = K_0 \exp[-\frac{E_a + \sigma \Delta V}{kT}] \quad (3)$$

with $k_0 = 1.72 \times 10^7 \text{Å}^2/\text{s}$ and $E_a = 1.3$ ev for dry oxidation.[2]

The stress profile $\sigma(x,X_{ox})$ across the oxide layer during growth is given by:[3,4]

$$\sigma(x,X_{ox}) = \sigma_{max} \exp\left[-\frac{1}{\tau}\int_{X_{ox}-x}^{X_{ox}} \frac{dt}{dX_{ox}}(u,X_{ox})du\right] \quad (4)$$

where σ_{max} is the maximum stress occurring at the interface ($x = 0$) arising from the difference of density between Si and SiO_2, τ is the Maxwellian relaxation time related to the viscosity η and the shear modulus μ by $\tau = \eta/\mu$. σ_{max} is known to be on the order of 10^{10} dyn/cm^2.[3-5]

The average stress σ_{av}, the total stress σ_t in the oxide layer resulting from oxidation are readily deduced after integration of the stress profile $\sigma(x,X_{ox})$.[3]

The refractive index n of the stressed oxide layer can be derived from the average stress by means of a first order compressibility relationship:[6]

$$n = n_0 + \frac{\Delta n}{\Delta \sigma} \sigma \quad (5)$$

where $\Delta n/\Delta s$ is the compressive coefficient (= 9×10^{-13} cm^2/dyn for SiO_2), N_0 is the refractive index of the oxide in free state (# 1.46 for SiO_2).

The corresponding density ρ of the oxide layer after oxidation can also be calculated using the Lorentz-Lorenz relation.

RESULTS AND DISCUSSION

The initial regime of dry thermal oxidation of silicon has been investigated using Rel. 1. It is clearly found that the existing stress profile in the growing oxide layer gives rise to a shift from the straight line of the Deal-Grove case in the dt/dX_{ox} vs

X_{ox} plots as already mentioned.[1,2] Table 1 shows a comparison between experimental data and theoretical values calculated using Rel. 2 with σ_{max} = 4×10^{10} dyn/cm^2 and the diffusivity ratios far from both and close to the interface.[4] The characteristic length over which this initial regime occurs, ranges from less than 5 nm at 1000°C to more than 65 nm at 750°C.[2]

	(100) Si				(111) Si	
T(°C)	780	870	893	980	880	930
D/Di(exp.)	>3	5.5	4.5	>1	3.2	3.9
D/Di(th.)	6.5	5.6	5.4	4.8	5.5	5.1

Table I. Ratio of the diffusivity in the oxide far from and close to the interface (after Ref. 4).

This results, throughout Rel. 3, in a change of the effective activation energy E_a of the parabolic constant for the same temperature range as can be seen in Fig. 1. Thereby E_a is about 1.3 eV above 900°C whereas it reaches 2-3 eV around 850°C.[4]

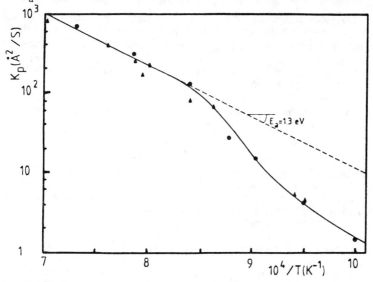

Fig. 1: Parabolic constant vs. reverse oxidation temperature (from Ref. 4).

Indeed, this behaviour is mainly indicative of a drastic stress dependence of the activation energy via the free energy (see Rels. 2-3) and more weakly of a change in the diffusion mechanism.[4]

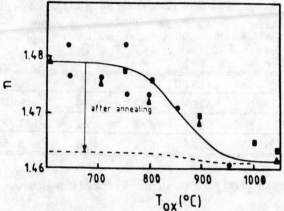

Fig. 2: Refractive index vs oxidation temperature for 100 nm thick SiO_2 layers (From Ref. 3).

Besides, the temperature dependence of the diverse stress levels in SiO_2 films have been favorably compared with experimental data.[3] For instance, as shown in Fig. 2, typical refractive index data vs oxidation temperature have been fitted using Rels. 4-5 with a maximum stress $\sigma_{max} = 3 \times 10^{10}$ dyn/cm^2.[3] This picture not only shows that the high values of refractive index for low temperature grown silica films are due to the presence of highly compressive oxidation stresses but that the stress relaxation process used by the stress state model in case of kinetic studies, holds also quite well for this particular application. Moreover, the stress values found independently in the kinetic analysis[4] and in the refractive index or density investigation[3] are rather close together.

CONCLUSION

As seen in this paper, the stress state model of oxidation is a useful tool for the modelling of the initial regime of dry thermal oxidation of silicon. Especially it provides a rather good description of the oxidation kinetics in the VLSI thickness range on a wide temperature scale. Even in this low thickness regime, the oxidation kinetics is likely based on a diffusion dominated process as recently inferred by Ng et al.[7]

Moreover, the model is also relevant for the evaluation of the stress levels within growing oxide layers.

Finally, it is worth noting the consistency of the stress values independently found from the kinetic point of view and the stress point of view.

REFERENCES

1. Fargeix, A., Ghibaudo, G. and Kamarinos, G., J. Appl. Phys., 54, 2878 (1983).
2. Fargeix, A. and Ghibaudo, G., J. Appl. Phys., 54, 7153 (1983).
3. Fargeix, A. and Ghibaudo, G., to be published in J. Phys. D. (1984).
4. Fargeix, A. and Ghibaudo, G., to be published in J. Appl. Phys. (1984).
5. EerNisse E.P., Appl. Phys. Lett., 35, 8 (1979).
6. Primak, W. and Post, G., J. Appl. Phys., 30, 779 (1959).
7. Ng, K., Polito, W. and Ligenza, J., Appl. Phys. Lett., 44, 626 (1984).

Part X. Design Automation, Device Layout and Packaging

PHYSICAL INTUITION AND VLSI PHYSICAL DESIGN

Ralph Linsker
IBM Research, Yorktown Heights, NY 10598

ABSTRACT

We characterize physical design of chips and packages as a set of constrained global near-optimization problems. Heuristics appropriate to these problems are discussed, including some that are strongly motivated by analogies to physical processes. We contrast several problem-solving approaches or frameworks: hierarchical decomposition, 'greedy' algorithms, probabilistic methods (e.g. Monte Carlo simulated annealing), and iterative-improvement methods using adjustable penalty functions. We emphasize that the discovery of a physically-appealing analogy can aid in the choice of an approach, but does not substitute for the use of problem domain-specific knowledge as well.

To illustrate issues guiding the choice of problem-solving framework and heuristics, we discuss an iterative-improvement interconnection routing system in production use at IBM. This system has yielded significant savings in manual effort and fabrication costs, and provided greater design flexibility while satisfying technology constraints, in a range of applications, compared with some earlier routers that have been used.

INTRODUCTION

This talk has two purposes: to discuss some physical-design issues and heuristic approaches that are relevant to a physics-oriented audience; and to describe a particular design system, based on some of these ideas, which has demonstrated practical advantages compared with some prior methods. In the first, more general, part of the talk, we outline some problems that arise in the physical design of computers, some physical analogies that have been found useful in developing design automation tools, and why physical analogies have proved relevant in this area. Several types of problem-solving approaches used in physical design are contrasted. We then discuss an iterative-improvement wiring design system in production use at IBM, emphasizing issues guiding the choice of heuristic approach, and results obtained using this system.

PHYSICAL DESIGN

Physical design (PD) comprises the specification of the detailed physical layout of computer components and wiring, given the list of connections that are required to be implemented between circuits. PD generally is divided into three steps:

- Partitioning -- the assignment of circuits to particular chips, of chips to first-level packages (modules or cards), etc., so as to satisfy constraints on the number of I/O's and circuits at each packaging level;

- Placement -- the arrangement of components on their respective next-level packages (e.g. circuits and macros on chips, chips on modules, etc.), so as to near-minimize connection lengths, meet timing constraints, etc.; and

- Wire routing -- the specification of the interconnection paths between components at each package level, so as to satisfy certain constraints (e.g. no wire crossings within a printed-circuit plane) and optimize performance considerations (e.g. wire length, crosstalk noise coupling between interconnections).

There is typically feedback between these steps, and simultaneous placement and wiring (or partitioning and placement) can be done.

ROLE OF PHYSICAL INTUITION AND ANALOGY

Physical analogy has motivated some detailed heuristic approaches in PD. These can involve the motion of objects (components, wires) in potentials tailored to the PD problem goals. An example is force-directed placement (attractive between components to be connected, with a repulsive component-overlap potential). One can similarly consider an approach to wire routing wherein each segment of a path between two components moves in a potential defined by the positions of other wires, terminal pins, package blockages, etc., or wherein an optimal path is found (by other means, e.g. a 'mazerouter' algorithm) for particular choices of potential (cost) function that are physically motivated.

Physical analogy has also proved fruitful at a higher heuristic level than that of the detailed move or potential-function definition. The Monte Carlo simulated annealing method of Gelatt and Kirkpatrick[1] is an example of this, in which the problem of trapping in high-lying local minima of a Hamiltonian during an optimization process is avoided by allowing 'uphill' moves with a probability determined by the change of energy for that move, and by a time-dependent 'temperature'; specifically, with probability $\exp(-\Delta E/T)$. When the Hamiltonian or objective function is such that the magnitude and direction of local energy changes provide a guide to directions of global configuration improvement, this method provides a means of searching an energetically favorable subset of configuration space (i.e. fruitfully limiting the search) without getting trapped in high-lying local wells.

What parts of the simulated annealing method are key to the solution of PD problems, and how does the method relate to other problem-solving approaches? For a PD problem, neither the particular choice of Hamiltonian nor the set of allowed moves need be taken as fixed. Since transformations of the Hamiltonian are allowed, the particular choice of the Boltzmann probability $\exp(-\Delta E/T)$ loses much of its force. We are left with a probabilistic mechanism for making uphill moves; one which initially (when T is large) distinguishes only gross differences in configuration energies, then (at lower T) distinguishes finer differences but loses the capability of escaping from too-deep local wells. A natural connection between this method and hierarchical methods thus appears. This connection suggests that, to save computing time, coarse moves and coarse evaluation of the Hamiltonian should be used when T is large. It also suggests that we can view simulated annealing within the context of other methods for making 'uphill' moves (i.e. backup within a heuristic search tree when one path has led to an inadequate solution). Finally, we note that there are many ways of controlling the course of an optimization process; whether control of a temperature T(t) is better than use of some other control strategy (or set of heuristics) is a problem domain-specific issue.

Why is there a link between physics and PD problem-solving? First, physical systems can be used as formal analogues to certain problems, as in the examples of downhill search in a specified potential, and of annealing. Is there a deeper connection, between the nature of the physicist's tools and certain classes of problems?

Physicists are used to thinking about how 'simple' systems respond to forces. The systems may be 'simple' by virtue of having few particles, or by virtue of being described by a Hamiltonian of simple form even though they may comprise a large number of particles. In the PD problems, the 'goodness' of a final configuration is described by a Hamiltonian containing many terms (e.g. thousands, one for each connection in a placement or partitioning problem), but with a basically simple structure (e.g. of the form $H = \Sigma J_{ij} f(x_i, x_j)$ where J_{ij} is 1 if components i and j are connected, and zero otherwise, and f is a function of the positions of the components, in a placement optimization problem). The structure of the J matrix is complicated and may appear random, but it is far from random -- it is determined by design decisions the nature of whose effect on J is still not well understood, although certain empirical relations (such as Rent's rule) are known. Nonetheless, if the structural complexity of the Hamiltonian is not critical to the optimization problem, we can treat it as a relatively simple Hamiltonian to which physical intuition (e.g. a spin glass analogy) can usefully be applied. If this complexity is critical, physical intuition or analysis may be of little help. Systems of complex organization -- biological or artificial -- are of this character. Methods of analysis and representation far removed from those used for the simpler physical systems -- e.g. representation of system function by sets of interacting rules rather than by a Hamiltonian, may be most appropriate (i.e. give the most insight, or facilitate system simulation).

We can look at this issue another way. For the PD problems, each of the constraints and penalty terms is local (involves only a few objects). The constraints are coupled; therefore a global solution is needed, rather than a patchwork of local solutions. But because each constraint is local, the problems lend themselves to the use of physical analogies (such as objects moving in a potential). Also, because the obvious objective functions increase in value gradually as one moves away from a good configuration using the obvious move set ('graceful degradation' of the solution), the local behavior of the objective function can be a useful guide to the location of deeper wells; hence simulated annealing can be effective. In harder problems, a non-obvious representation may need to be defined, in order to achieve graceful degradation. If several such representations, perhaps hierarchically organized, are required, then the plan defined in terms of these representations becomes the dominant element of the solution process, and the various objective functions take on a mere scoring role. [Example: writing a program (or building a machine) to do a task. An objective function at the level of compiler or machine language would have a very large value, and be useless for guidance, for almost all configurations -- almost all sequences of code that would be reached by the move generator are equally worthless. A plan embodied as a hierarchy of flow charts at different levels of detail, concluding in the compiler language code, can in principle be evaluated much more meaningfully for its closeness to a correct or optimal plan (at each level).]

'Hard' problems don't have a readily definable Hamiltonian that guides the system toward a good configuration. What happens to the Hamiltonian as we pass from easier to more difficult problems? Consider an analogy between such problems and a physical system with external interactions, in which our goal is to use the external interaction as a controller for driving the physical system toward a preferred state. We can consider several processes for accomplishing this, ordered by increasing complexity of the external interaction.

- An annealing method, in which the system is placed in contact with a heat bath of controlled temperature $T(t)$.

- A programmed sequence of process steps to which the system is subjected, without feedback (an external time-dependent driving force).

- As above, but with a few (e.g. thermodynamic) system variables influencing the process through a feedback loop.

- A very large number of system variables influencing the process through feedback. The interaction may be highly nonlocal. (We can equally well think of the 'system' as comprising the controller in this case, since the coupling is strong and complex.) The earlier comments concerning systems of 'biological' complexity start to apply in this case.

In the next section we discuss several types of PD heuristic approaches, which tend to fall within the first three of these process categories.

HEURISTIC APPROACHES IN PHYSICAL DESIGN

In each phase of physical design, the problem is to find a configuration that is near-optimized with respect to a specified objective function ('OF') and possibly subject to a set of constraints. The value of the OF typically depends upon a large amount of information about the configuration, sometimes in a tightly coupled way. The algorithms invoked at each step of the optimization process generally treat only a small amount of information at a time, and/or decouple portions of the problem (e.g. by routing the connections one at a time); the 'field of view' (FOV) of each algorithm is limited. A plan is required to give overall direction to the solution process, such that one can work with limited-FOV algorithms yet converge to a globally near-optimal solution. Several types of plans have been used.

- Hierarchical. One can group the elements (components, wires, etc.) into initially coarse, then progressively refined, categories. The FOV of each algorithm can thereby be relatively small in terms of information content, even though the algorithm may be manipulating a large number of elements (as blocks) at each of the early steps. A method due to Burstein et al[2] is an example of this approach, in which wiring is performed by solving a succession of constrained flow problems.

- Greedy algorithms. Methods of this type locally optimize at each step, but favor earlier- over later-wired (or placed) elements. The result tends to be a far from globally optimal solution, and the 'greedy' phase is often followed by use of highly problem-specific repair heuristics. Clustering methods for placement, and many sequential wire routing methods, are examples of this approach.

- Basic iterative improvement. The OF is fixed and a series of iterative passes is performed. At each pass, a sequence of trial local moves is considered. Each move is accepted if it results in an improvement (reduction) of the value of the OF. This typically yields a solution that is locally optimal against any move, but far from a global minimum of the OF.

- Controlled 'uphill' moves. Here the convergence to a high-lying metastable configuration (as in the previous method) is avoided by accepting certain moves that increase the OF. The goal is to accept such moves in a way that selectively favors the exploration of lower-lying wells with eventual convergence to a globally near-minimal configuration. The simulated annealing method is such a method that is strongly motivated by a physical analogy between a constrained optimization problem and a statistical-mechanical system with a Hamiltonian characterized by frustration.

- Iterative improvement with variable penalty functions. Although the 'goodness' of an eventual solution is defined by the OF and by specified constraints, there is no reason to limit oneself to the OF and constraints when deciding the worth of intermediate configurations or the 'legality' of moves. The additional freedom provided by this approach has several advantages. Varying the penalty function with each iterative pass allows one to embody problem-specific planning heuristics (e.g., weights assigned to various desiderata may be changed during the course of the optimization). A different penalty function may be applied to each of several classes of objects (e.g. different types of connections), to achieve design goals and/or to embody additional problem-specific heuristics. The process freedom is essentially at the level described at the end of the previous section as that of the controller with feedback of a small number of system variables. Furthermore, because the effective objective function is time-varying, the locations of the local minima change, allowing (though not guaranteeing) the system to avoid becoming trapped in local wells. This is particularly useful in cases, such as that of wire routing with paths of arbitrary shape, in which the time required to compute the effects of a single move (a rip-out and reroute of a single wire) is great enough that simulated annealing would be far too slow.

In the next section we turn from this general discussion of heuristic approaches, to a specific example of a wiring design system based on an iterative-improvement variable penalty function approach. We illustrate the flexibility that is provided by such an approach, and compare the results obtained in production and test applications with those of sequential routers that have been used.

EXAMPLE OF AN ITERATIVE-IMPROVEMENT DESIGN SYSTEM

We describe briefly some features of a system (VIKING) based on an iterative-improvement variable penalty function driven method, which is in production use for routing of interconnections on cards and boards at IBM. A more detailed discussion is given in Ref. 3.

VIKING comprises two phases, an optional global (or coarse) router and a detailed (or exact) router. The global router assigns a preferred routing on a coarse grid, with a preferred layer (or plane-pair) assignment when appropriate, to each connection. This is done by making an initial assignment of a path to each connection, then performing a sequence of iterative passes. In each pass, the connections are, one at a time, removed and rerouted so as to minimize a prescribed penalty function. The basic penalty terms are for wire length and congestion (wiring load in relation to track capacity, at each point on the global grid). Other terms penalize e.g. global path bends and routing through specified regions; special length and other routing constraints for specified subsets of connections can also be handled. Once a satisfactory global routing is achieved, the preferred global paths are passed as input to the detailed router.

In contrast to sequential routers, VIKING starts the detailed routing by assigning paths to all connections.[4] (If a global routing has been done first, the global path assignments are respected at this stage.) The initial detailed routing is in general highly 'illegal' (e.g., containing intersections within a printed-interconnection plane). An iterative refinement of the paths is then performed, in a manner similar to the global router. Typical penalties can include (depending upon the application) those for wire length, intra-plane intersections (which represent shorts in a printed-circuit technology), bends, vias (through-hole connections between planes), excessive wire adjacencies (related to electrical crosstalk noise), and other occurrences affecting cost, manufacturing process complexity, and reliability. Routing of discrete (as opposed to printed) wires, and eight-way (including diagonal) routing, are also available with the system.

We note several points of contrast between this approach and conventional sequential routers. The latter use a 'greedy' algorithm, routing each connection in turn if legally possible, and setting aside routing failures for later automatic or manual handling.

- By treating the usual 'illegalities,' such as intersections, as occurrences to be penalized rather than prohibited, VIKING allows design tradeoffs not available with sequential routers.

- In multiplane packages with orthogonal routing, each plane is conventionally assigned to carry wiring predominantly in one or the other direction (x or y). This is occasionally required by the particular packaging technology, but is often a condition imposed in order to aid the wiring design process by avoiding situations in which pin escape or routing of late-wired connections would be blocked if x and y wiring were freely allowed in each plane. This can pose a particular problem if available vias are sparse or if used vias block wiring channels. In VIKING there are no 'early' or 'late' wired connections, and the usual prohibition against 'free-form' or extensive 'wrong-way' wiring (WWW) can be dropped. Dropping this constraint allows one to reduce the number of used vias, and overall wire length, below that otherwise required. Even where free-form or extensive WWW is undesired or unnecessary, the WWW routing flexibility provided by VIKING has been exploited to enable wiring escapes from congested regions, in fewer wiring planes than would be needed if WWW were prohibited. This reduction in the number of wiring planes can contribute to substantial fabrication cost savings.

- As a useful by-product of the iterative approach, wires with residual illegalities remain on the board at the end of a VIKING run. These can be highlighted and usually manually revised locally (by moving paths in the immediate vicinity of the illegality), rather than having to find a complete manual routing for each unrouted connection, against a congested background, as is the case with sequential routers. This is especially useful in large card and board applications; the quality of the resulting routing is improved, manual effort per illegality is reduced, and tortuous paths that can lead to excessive delays in high-performance applications can be eliminated or better controlled.

- As stated earlier, inter-pass variation of the form of the penalty function not only provides a way to incorporate useful heuristics into the routing process, but aids escape from high-lying local minima, which is a problem with 'greedy' and with basic (fixed penalty function) iterative-improvement algorithms.

- The iterative-improvement approach described also allows us to extend the range of applicability of a class of standard methods known as 'mazerouter' algorithms.[5] These methods perform a minimum-cost routing of a single connection (taking into account the rest of the board layout). Such algorithms require that the cost function for a path be linear in the number of occurrences (on that path) for each penalty type. This limitation is undesirable for several reasons. For example, one may want to impose steep penalties for path lengths violating maximum or minimum length constraints; or one may need to inhibit inter-path adjacencies exceeding a threshold length. An iterative-improvement variable penalty function method can be used to eliminate this linear cost function limitation. This is done by iteratively rerouting a path using a penalty function with coefficients that themselves depend upon the prior history of the routing of that connection. This provides an improvement in the basic one-connection router algorithm, independent of the global-optimization and other issues relevant to the multi-connection routing problem. References 3 and 6 give further details concerning this point.

Some results obtained using VIKING are summarized below. Illustrations will be shown during the talk, and further details are given in Ref. 3.

- In printed-circuit card applications having 1000-2000 connections and 5-8 wiring planes, with sparse vias, VIKING decreased excess over 'minimum-Manhattan' wiring length (i.e. decreased total detour length) by factors of 1.5 to 3.5 (compared with a sequential router used within IBM). Via count was decreased by about 20%, and manual embedding effort was decreased by a factor of 2-3 and sometimes much more. Substantial fabrication savings can result from reductions in the number of wiring planes required by VIKING, and from package simplifications that become possible (e.g. reduction in the number of via sites that must be provided).

- For the first production case on which a VIKING run was attempted, a card with 1612 connections and six wiring planes had been wired by a sequential router with 148 remaining unrouted connections, which required three weeks to embed manually. VIKING wired this card with no unrouted connections, and with 25 crossings requiring an estimated one day for local manual revision. In this example, VIKING's superior performance was due to a combination of two factors: use of iterative-improvement rather than sequential-routing methods, and reduced via requirements (yielding improved wireability in this via-sparse case) due to use of 'free-form' or extensive 'wrong-way' wiring.

- VIKING provides efficient routing in directional constrained wiring planes, as well as for 'free-form' (unconstrained) problems, and does not require a separate channel router to handle dense wiring in via-less regions. This will be illustrated in the talk.

- A printed-circuit board containing about 8000 connections and many congested regions was routed using VIKING in six wiring planes, compared with the eight-plane design obtained using another (sequential) production router. Salient features include:

  - Incorporation of threshold adjacency control in the VIKING run produced a routing that contained 25 times fewer violations of a particular intra-plane crosstalk criterion than were generated using the sequential router, even though the wiring density per plane was over 30% greater in the VIKING case. This greatly aids in the subsequent manual revisions needed to satisfy detailed crosstalk constraints. Furthermore, the use of dynamic adjacency control also avoids the impairment of wireability the results from artificially blocking wiring channels (as is conventionally done to control crosstalk).

  - Some of the congested regions had too many wiring 'escapes' to be routable using the wiring capacity of three 'x' and three 'y' planes. Even though directional preferences were prescribed for each plane in this case, VIKING was able to circumvent this escape limitation by an appropriate choice of penalty functions and by means of controlled WWW in the congested regions.

  - Automatic control of manufacturing cost and reliability factors during the wiring process is provided by the VIKING approach.

- For cases containing extensive blockages in one or more planes, or regional preponderances of x- over y-oriented connections, the free-form wiring capability allows greater wireability than that obtained using directionally committed planes. An example illustrates how VIKING 'discovers' unaided a useful heuristic for organizing wiring directionality; this example is suggestive of some aspects of chip wiring with blockages caused by 'macro' placement.

SUMMARY AND CONCLUDING COMMENTS

Physical design, viewed as a class of constrained global near-optimization problems, has characteristics that allow the fruitful application of physical intuition and explicit physical models in their solution. Some overall approaches used in PD, both physically motivated and otherwise, have been discussed. An example of a flexible iterative-improvement wire routing system, and the advantages it provides over some other approaches, was presented.

Complex problems require more sophisticated sets of heuristics. Viewed in terms of the complexity of the heuristic set (as opposed to the effort that may be involved in implementing the algorithms), the usual approaches in automated PD have rather little heuristic structure. Specification of a particular objective function (if only one is to be used during the optimization process) can be thought of as constituting just one heuristic; specification of a simulated annealing control strategy as constituting one more (if the particular annealing schedule T(t) is included). Methods involving a sequence of penalty functions, with or without a controlled-uphill method included, introduce a few more heuristics, particularly if there is feedback influencing the penalty-function schedule. This is a far cry from the complexity of a large interacting set of heuristics, such as is found in certain expert systems and may be found with a still greater degree of complexity in future 'learning' systems. This relative simplicity may reflect a relative structural simplicity in the PD problems themselves (as they have been defined thus far), but may also point to opportunities for substantial future invention in these and other areas of the design of computers and other complex systems.

REFERENCES

1. S. Kirkpatrick, C. D. Gelatt, Jr., and M. P. Vecchi, Science 220, 671 (1983); S. Kirkpatrick, J. Statist. Phys. 34, 975 (1984).

2. M. Burstein, S. J. Hong, and R. Pelavin, Proc. VLSI-83, p. 45 (Trondheim, 16-19 August 1983).

3. R. Linsker, IBM J. of Res. and Develop. 27, Sept. 1984 (in press). The VIKING system was developed initially by the author, then implemented for production use within IBM in work with J. F. Cooper and C. N. Lamendola.

4. Independent work on iterative-improvement routers has been described by A. Moore and C. Ravitz, IBM Tech. Discl. Bull. 25 (7B), 3619 (1982); and earlier by F. Rubin, 11th Design Automation Workshop, p. 308 (1974).

5. For an early discussion, see C. Y. Lee, IRE Trans. on Electronic Computers, EC-10, pp. 316-365 (1961).

6. R. Linsker, IBM Tech. Discl. Bull. 27, No. 1A, 399-406 (1984).

CONCEPTS OF SCALE IN SIMULATED ANNEALING

Steve R. White
IBM Thomas J. Watson Research Center, Yorktown Heights, NY 10598

Abstract

Simulated annealing is a powerful technique for finding near-optimal solutions to NP-complete combinatorial optimization problems. In this technique, the states of a physical system are generalized to states of a system being optimized, the physical energy is generalized to the function being minimized, and the temperature is generalized to a control parameter for the optimization process. Wire length minimization in circuit placement is used as an example to show how ideas from statistical physics can elucidate the annealing process. The mean of the distribution of states in energy is a maximum energy scale of the system, its standard deviation defines the maximum temperature scale, and the minimum change in energy defines the minimum temperature scale. These temperature scales tell us where to begin and end an annealing schedule. The "size" of a class of moves within the state space of the system is defined as the average change in the energy induced by moves of that class. These move scales are related to the characteristic temperature scales of a system, and show that a move class should be used when it gives an average change in energy on the order of the temperature. This, in turn, helps improve the performance of the algorithm.

Introduction

Many problems encountered in VLSI design are NP-complete combinatorial optimization problems. Briefly, this means that globally optimal solutions to these problems cannot be found because the computing time required to do so grows exponentially with the problem size. Hence, heuristic methods must be used, which find solutions that are close to optimal in a tractable amount of time.

A good way to find low-energy states of complex physical systems such as solids is to heat the system up to some high temperature, then cool it slowly. This annealing process lets the system settle into regions of low energy, while not getting trapped in higher-lying local minima. Simulated annealing is a heuristic optimization technique that uses methods from statistical physics to find near-optimal solutions to combinatorial optimization problems in much this same way.[1]

In abstracting this method to optimization problems, the objective function of the system to be optimized is identified with the energy in the physical problem. New states of the system are generated by applying a set of moves to the system, and accepting or rejecting the moves based on the Metropolis criterion:[2]

If $\Delta E \leq 0$, accept the move;
If $\Delta E > 0$, accept the move with   (1)
   probability $P(\Delta E) = e^{-\Delta E/T}$;

where $\Delta E \equiv E_{final} - E_{initial}$ and $T \equiv$ Temperature.

As these moves are accepted or rejected at a fixed value of T, the system tends towards thermal equilibrium at that temperature. In thermal equilibrium, the probability that the system is in a particular state with energy E is proportional to $e^{-E/T}$.

Characterizing <E(T)>

A typical annealing algorithm proceeds by starting the system at some high temperature. The system is allowed to approach equilibrium, at which time the temperature is reduced, the system allowed to equilibrate again, and so on. The process is stopped at a temperature low enough that no more useful improvement can be expected. This protocol for cooling the system is known as the annealing schedule.

In discussing annealing schedules, it is useful to graph $<E(T)>$, the average energy at a fixed temperature vs. the temperature. This graph will be referred to as the annealing curve. It tracks the progress of the annealing in a way that is tied to the fundamental quantities in the system, and is relatively independent of the details of how the annealing schedule is carried out. It allows us to examine empirically the relationship between characteristic scales in temperature and those in energy. The more usual graphs show energy vs. the number of moves attempted thus far. This is not as useful, since there are no characteristic scales in the number of moves, and because it does not reflect the temperature of the system at any point.

Two basic questions about the annealing schedule will be dealt with here. First, how high must the system be heated before annealing can begin? Second, how low must the system be cooled to achieve a good result? Useful answers to these questions will be as independent of the problem, and its size, as possible.

The problem used here for demonstration is that of minimizing the Manhattan lengths of wires in a circuit placement problem.[1] A circuit is thought of as a square object on a checkerboard. Each wire net connects some subset of the circuits, and the wires are constrained to be piecewise in the x or y directions. A basic move in this system consists of interchanging two adjacent circuits. The energy of the system is given by:

$$E \equiv \sum_{n \in \{\textbf{nets}\}} [l_x(n) + l_y(n)] \qquad (2)$$

where $l_x(n)$ and $l_y(n)$ are the x and y lengths of net n, respectively. This example was chosen because annealing is known to yield excellent solutions, and because it is typical of many optimization problems encountered in VLSI design. As will be seen, many of the results of this analysis are independent of the particular problem chosen.

Figure 1 shows a typical annealing curve for this problem. This example consists of 105 circuits and 85 nets, placed in a 10×11 grid. Each dot represents an average over a statistically large number of states of the system generated at a fixed temperature. Many of the problems that have been studied with annealing show similar annealing curves, and each has similar overall features:

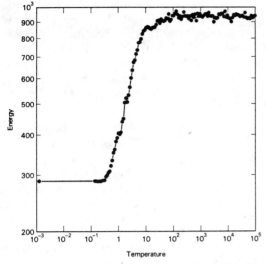

Figure 1: Energy vs. Temperature for a Typical Problem

(1) There is a minimum energy, E_0, below which the system never goes. This must be the case; the system has some finite global minimum.

(2) There is a maximum average energy, E_∞. There certainly must be an energy maximum; the system is finite, and some configuration provides a finite upper bound on the energy. Beyond this, in the infinite-temperature limit, any attempted move will be accepted, regardless of whether it increases or decreases the energy. Hence, at very large temperatures, the system moves randomly through its states, and the observed average energy is just the average of the energies of all of the possible states of the system.

(3) The annealing curve makes a transition at a low temperature, T_0, from an intermediate-energy regime to a low-energy regime, with $<E(T)>$ decreasing with $T > T_0$, and $<E(T)> \simeq E_0$ for $T < T_0$.

(4) The annealing curve makes a transition at a high temperature, T_∞, from the high-energy regime to the intermediate-energy regime. This transition is very broad.

The Density Of States

The density of states of the system, $\omega(E)$, contains all of the information pertaining to equilibrium properties of the system at all temperatures. Thus, it is an important tool for analyzing the behavior of annealing algorithms.

The density of states can be found empirically for any system by collecting the number of states with energy E as random changes are made to the system. In a simulated annealing program, this is equivalent to collecting the number of states with energy E while the temperature is held extremely high.

Figure 2 shows the density of states for wire length minimization, collected in just this way. Several features are worth noting. If there are N circuits in the problem, the total number of states (under the curve) is $N!$. In finding a near-optimal solution, we are primarily interested in only a few states around E_0 (the minimum energy), and these states are far to the left of the graph in Figure 2.

For $E \rightarrow E_0$, there are only a tiny number of states. This just says that the problem is a difficult one, much the same as that of finding the ground state of a

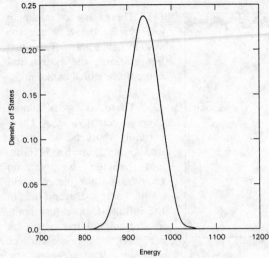

Figure 2: Density of States for a Typical Problem

collection of atoms. For $E \to \infty$, there are no states. This just says that the system is finite. The density of states is sharply peaked around $E = \bar{E}$, and this peak has a characteristic width in energy.

Annealing Limits as Characteristic Scales

For energies E near $\bar{E}$, the variable E may be taken to be continuous, to a good approximation. The equilibrium value of E at a temperature T is then given by:

$$<E> = \frac{\int_{E_0}^{\infty} dE \, E \, \omega(E) \, e^{-E/T}}{\int_{E_0}^{\infty} dE \, \omega(E) \, e^{-E/T}}. \quad (3)$$

For E "near" $\bar{E}$ (the average energy), we can take, to a good approximation,

$$\omega(E) \propto e^{-(E-\bar{E})^2/2\Delta^2}. \quad (4)$$

That is, the density of states is approximately Gaussian near the average energy.[3] When $\omega(E)$ is given by Equation (4), Equation (3) becomes

$$<E> = E_0 + \sqrt{2}\,\Delta \left[y + \frac{1}{\sqrt{\pi}} \frac{e^{-y^2}}{1 + \mathrm{erf}(y)} \right] \quad (5)$$

$$\text{where } y \equiv \frac{1}{\sqrt{2}} \left[\frac{\bar{E} - E_0}{\Delta} - \frac{\Delta}{T} \right].$$

The high-temperature limit is found by taking $|y| \gg 1$ and $y > 0$. In this limit,

$$<E> \simeq \bar{E} - \frac{\Delta^2}{T}. \quad (6)$$

The characteristic high temperature T_∞ can be identified with the temperature at which $<E>$ is just within thermal noise of $\bar{E}$. This is the temperature at which $(\bar{E} - <E>)/\Delta = 1$, that is, $T = \Delta$. Thus, if $T \gg \Delta$, the system is well within thermal noise of its infinite-temperature equilibrium condition. The characteristic temperature, above which raising the temperature does not affect the system significantly, is

$$T_\infty = \Delta. \tag{7}$$

These results characterize the annealing curve at the large energy and temperature scales.

The criterion for the system being "hot enough" can be rephrased by noticing that

$$<E^2(T)> - <E(T)>^2 = T^2 \frac{\partial <E>}{\partial T}. \tag{8}$$

Using Equation (6), this becomes

$$<E^2(T)> - <E(T)>^2 \simeq \Delta^2. \tag{9}$$

So the condition that $\frac{\Delta}{T} \ll 1$ becomes

$$\frac{\sqrt{<E^2(T)> - <E(T)>^2}}{T} \ll 1. \tag{10}$$

This can be monitored automatically in a program to determine when the system is "hot enough."

The low-temperature limit can be examined in a simple way. If a move changes the energy at all, the lower bound on this change is typically nonzero. In wire length minimization, the smallest nonzero change in the wire length is one. Consider a state with energy E_1 that is just above a nondegenerate local minimum with energy E_0. Suppose, for simplicity, that there is exactly one move that will take the system from E_1 to E_0. Let M be the total number of moves originating from the state E_1. (For the placement problem, $M \simeq 2L^2$.) Thus there are M moves, but only one of them goes downhill to E_0. To be in equilibrium with $<E> \le E_1$, we must have $e^{-(E_1 - E_0)/T} \le 1/M$. That is, $T \le (E_1 - E_0)/\ln N$. We can therefore identify the characteristic low temperature as

$$T_0 \equiv \frac{E_1 - E_0}{\ln M}. \tag{11}$$

The important point here is that, in this model, the qualitative behavior of $<E(T)>$ is completely characterized by four quantities: E_0, $T_0 \equiv (E_1 - E_0)/\ln M$, $E_\infty \equiv \overline{E}$, and $T_\infty \equiv \Delta$. There are no other interesting energy or temperature scales in the problem.

Scaling of the Annealing Limits

In many problems, the scaling of the temperature and energy scales with increasing problem size can be determined analytically. Consider the example of wire length minimization in a placement problem with N objects on a grid of $L \times L$

squares, connected by $\sim N$ wires. Since the smallest possible change in the total wire length as a result of moving a circuit is one, we find:

$$T_0 = \frac{E_1 - E_0}{\ln M} \simeq \frac{1}{2 \ln L}. \qquad (12)$$

A theoretical estimate of the minimum energy is nontrivial. Currently, the best such estimates are based on "wirability theory",[4] which is based on a scaling law for wiring topologies in VLSI design called Rent's rule. Wirability theory gives us the following for the minimum energy:

$$E_0 \sim NL^r \text{ where } r \simeq 1/3. \qquad (13)$$

The high-temperature behavior can be modeled as that of a system of $\sim N$ nets, randomly distributed on the grid. The distribution of energies is that of an N-step random walk with step size $\sim L$, so:

$$T_\infty = \Delta \sim \sqrt{N} L. \qquad (14)$$

Since an average random net has length $\sim L$,

$$E_\infty = \overline{E} \sim NL. \qquad (15)$$

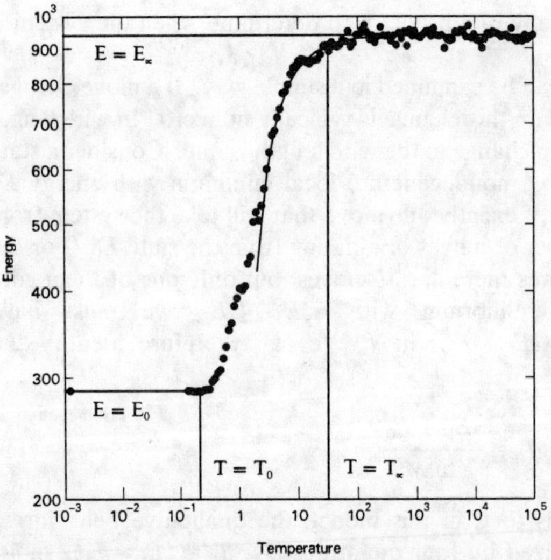

The annealing curve in Figure 1 is for a placement problem with 105 objects and 85 nets, on a 10×11 grid. The annealing scales for this problem are thus:

$$\begin{array}{c} T_0 \simeq 0.22 \\ E_0 \sim 226 \\ T_\infty \sim 102 \\ E_\infty \sim 1050. \end{array} \qquad (16)$$

Figure 3: Theoretical Fit to $<E(T)>$

Figure 3 shows the annealing curve from Figure 1, with the high-temperature data fitted to Equation (6), using the values $\overline{E} = 940$ and $\Delta = 32$. The value of $T_0 = 0.22$ is shown, and the system reaches a minimum of $E_0 = 287$. These values are in good agreement with the order-of-magnitude estimates of Equations (16), and they are in very good agreement with the values obtained by examining Figure 2. The fit to the high-temperature behavior of $<E(T)>$ is very good.

Comparing Annealing Limits

We can now examine the consequences of using annealing limits other than those derived above. Figure 4 shows three annealing curves for successive runs of the same placement problem, in which the system was heated up until $T \gg T_\infty$ then cooled down until $T \leq T_0$. The average solution is good, and there is a small spread among the solutions.

Figure 5 shows three annealing curves for this same problem, in which the system was not heated up to $T = T_\infty$. This is a poorer annealing schedule; the average solution is not as good, and there is a larger spread among the solutions. This occurs because the system never diffused out of a large-scale valley. This corresponds to not melting a solid completely before beginning to anneal it. Large-scale non-optimal features remain frozen into the system.

In this particular problem, this lack of large-scale diffusion could have been overcome by sitting longer at each temperature, and allowing the system more time to diffuse. As the starting temperature decreases, this technique becomes increasingly time-consuming, and eventually prohibitively so. The critical starting temperature at which it becomes prohibitive is likely to be problem-dependent, so the safest route is always to heat up until $T \gg T_\infty$.

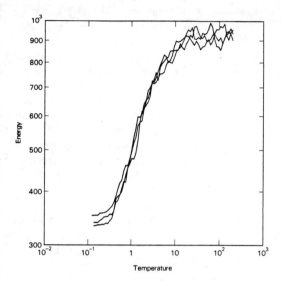

Figure 4: Good Annealing Curves

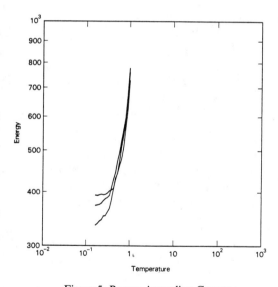

Figure 5: Poorer Annealing Curves

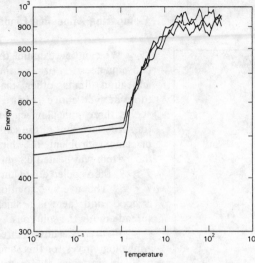

Figure 6: Terrible Annealing Curves

Figure 6 shows three annealing curves for this same problem, in which the system was not cooled down uniformly to $T = T_0$. Instead, annealing was terminated at $T \simeq 1$, and the system was quenched to very low temperature to allow it to go downhill into a nearby local minimum. The result is that the system became stuck in a high-lying local minimum, with no chance of getting out. This corresponds to quenching during annealing, which freezes in random small-scale features. The average solution is very poor, and the spread among the solutions is large. Unlike the previous case, only a prohibitive amount of low-temperature diffusion would allow the system to escape from its local minimum. Thus it is imperative to allow the system to cool until $T \leq T_0$.

Move Classes and Move Scales

There are also characteristic length scales in many optimization problems. By connecting length scales with temperature scales, it is possible to use the concept of the length of a move to enhance the performance of the annealing algorithm. To do this, consider a system with a topology which is *defined* by a set of moves. Call these *moves of length 1*. These moves take the system, by definition, to adjacent states in the space.

Now, define *moves of length ℓ* as moves connecting states that were previously connected by making at least ℓ moves of length 1. This simply increases the connectivity of the space. A single move of length ℓ can often be made at less computational cost than ℓ moves of length 1. Furthermore, "larger" moves take you "farther" in the original space, so larger moves can help the system approach equilibrium faster.

For example, consider N objects on an $L \times L$ grid at infinite temperature. With moves of length 1, moving a single object to an adjacent grid location, the equilibration time is $\sim NL^2$. With moves of length L, the equilibration time is reduced to $\sim N$.

Let ΔE be the change in the energy as the result of a move. For annealing to work well, the move set should be defined so that $<|\Delta E|>$ increases monotonically with ℓ. Figure 7 shows the average change in energy as a function of move length for the placement example. This demonstrates that this definition of move length is well suited to the placement problem.

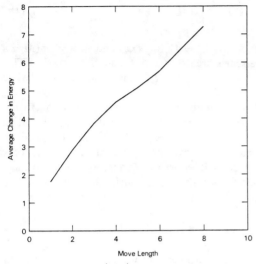

Figure 7: $<|\Delta E|>$ vs. Move Length

There is a direct connection between temperature scales and move scales. At T_∞, most *large* moves ($\ell \sim L$), and essentially all small moves, are accepted by the Metropolis criterion. At T_0, few *small* moves ($\ell \sim 1$), and essentially no large moves, are accepted.[5] The fact that there are no other interesting temperature scales in the problem implies that there are no other interesting length scales.

The notion of move scales can be generalized to arbitrary sets of moves. As with moves of different lengths, the criterion for good move sets is that each set should yield a characteristic value for $<|\Delta E|>$, and these values should be well separated from each other.

This suggests a strategy for adapting the move set being used to the temperature scale. At the highest temperatures, make only the largest moves (i.e. the ones with the largest $<|\Delta E|>$). Smaller moves change E by essentially trivial amounts, and are not useful in helping the system come to equilibrium. As the temperature decreases, use smaller moves (i.e. ones with smaller $<|\Delta E|>$), appropriate to that temperature. Larger moves change E by such a large amount, that they are accepted only with an extremely small probability.

The final consideration in choosing a move class for a temperature is to note that the change in E must be fairly continuous as it proceeds to equilibrium at that temperature. Thus, each move in the move class should change E by somewhat less than the average thermal fluctuations at equilibrium. This means that, at any temperature T, move classes should be selected which give

$$<|\Delta E|> \simeq T, \qquad (17)$$

to within a standard deviation in the distribution of $|\Delta E|$'s, with the constraint that

$$<|\Delta E|> \; < \; \sqrt{<E(T)>^2 - <E^2(T)>} \qquad (18)$$

Problems with More Than Two Scales

In the placement problem considered above, there were only two characteristic temperatures, and two corresponding characteristic energies. It is entirely possible for a problem to have more scales. Consider a placement problem in which the y-contribution of the nets is weighted much more than the x-contribution:

$$E \equiv \sum_{n \in \{\text{nets}\}} [l_x(n) + \gamma l_y(n)] \ ; \ \gamma \gg 1. \tag{19}$$

There are now two low temperature scales: $T_0^x = 1/\ln(M/2)$ and $T_0^y = \gamma/\ln(M/2)$. Similarly, the two high temperature scales are: $T_\infty^x \sim \sqrt{N L}$ and $T_\infty^y \sim \gamma \sqrt{N L}$. This system is "hot enough" when $T \gg \max(T_\infty^x, T_\infty^y) = T_\infty^y$. It is "cool enough" when $T \leq \min(T_0^x, T_0^y) = T_0^x$.

In such systems, it is useful to separate the move classes according to the different temperature scales. In this case, moves in the x and y directions should be put into distinct classes. For $T > T_\infty^x$, only moves in the y direction need be considered. For $T_0^y < T < T_\infty^x$, both classes of moves must be used. For $T < T_0^y$, only moves in the x direction need be considered. For $T < T_0^y$, limiting moves to the x direction only can allow the algorithm to use only half as much time as it would have if it considered moves in both directions.

Conclusion

For many combinatorial optimization problems to which simulated annealing can be applied, an examination of the density of states shows that their annealing properties can be characterized in terms of four energy scales: the minimum energy, the minimum change in energy near E_0, the average energy, and the width of the distribution of states. These scales allow us to know when the system is "hot enough", and when it is "cool enough" during annealing.

Move sets should be defined so that the average change in energy induced by a set increases monotonically with the "length" of the moves. Using mostly moves for which the average change in energy is approximately equal to the temperature automatically adapts the move set to the temperature scale.

References

[1]   S. Kirkpatrick, C.D. Gelatt, Jr., M.P. Vecchi, *Science* **220**, 671 (1983)

[2]   N. Metropolis, A. Rosenbluth, M. Rosenbluth, A. Teller, E. Teller, *J. Chem. Phys.* **21**, 1807 (1953)

[3]   The density of states near $\overline{E}$ can be shown to be Gaussian for wire lengths in placement, under the approximation that the locations of the endpoints of the wires are uncorrelated for these states. In other problems, the density of states is not Gaussian. In many of these, however, it is sharply peaked at some energy $\overline{E}$, and the distribution has a characteristic width Δ. Thus a Gaussian distribution captures essential qualitative characteristics of a wide class of problems.

[4]   Wilm E. Donath, *IEEE Transactions on Circuits and Systems* **26**, 272 (1979)

[5]   This is in contradiction to the commonly-held view that the system is "hot enough" when most moves (of an arbitrary kind) are accepted, and "cool enough" when few moves (of an arbitrary kind) are accepted.

Part XI.   III-V Compounds

HIGH SPEED GALLIUM ARSENIDE TRANSISTORS FOR LOGIC APPLICATIONS

Lester F. Eastman
School of Electrical Engineering, Cornell University
Ithaca, NY 14853

ABSTRACT

Compound semiconductors like gallium arsenide yield high switching speeds for logic applications. Velocity saturation values of 1.2 and 1.8 x 10^7cm/s at room temperature occur in GaAs for ordinary doped-channel MESFETs and modulation-doped heterojunction FET's respectively. Sub-half-micron GaAs devices are shown to have 4.0 and 8.0 x 10^7cm/s velocity values for gradual acceleration and quick acceleration respectively. These high velocity values, yielding high g_m values, are possible with low voltage swings. These allow very high switching speeds. To date 15 psec switching with doped-channel MESFETs, and 12 psec with modulation-doped FET's have been obtained with unity fan out at room temperature respectively. The latter device is expected to switch in less than 5 psec for shorter drift distances. The physical electronics of the various structures, and some the technologies used for growing and processing devices will be presented.

INTRODUCTION

Compound semiconductors, their alloys, and their heterojunctions in electron devices have the capability of yielding high electron velocity at low voltages. The electron mobility is high in gaAs, and electron velocity saturates at a value of 1.2 x 10^7cm/s, for devices over one micron in length. Submicron devices, especially with lighter doping densities, yield even faster electrons, approaching the ballistic limit where electron kinetic energy nears that of the potential drop. With gradual acceleration, average velocity during transit time for sub-half-micron GaAs devices can exceed 4 x 10^7cm/s. For quick aceleration over a .30 V potential step this average electron velocity in sub-half-micron GaAs devices can exceed 8 x 10^7cm/s. With transit times in the range of under 1 picosecond in the future, transistor f_T values exceeding 150 GHz can be expected.

Such high average velocities in FET devices yield high g_m values and g_m/C_{in} ratios even at moderate device voltage values. C_{in} is the device input capacitance and g_m is its transconductance. Because of the high g_m/C_{in} and moderate switching voltage values above the noise margin limit, very high speed FET logic is being developed. By using the difference in band gap across an Al,GaAs/GaAs emitter-base junction, high speed heterojunction bipolar transistors are also possible. High current gains, with high base doping and lower emitter doping are possible, lowering parasitic circuit element values. In addition, electrons can travel nearly ballistically through the base, lowering the base transit time by more than 10:1.

In addition to high speed transistors, compound semiconductors can be used to detect and generate short optical pulses. Thus communication from computer to computer, or even chip to chip, can be easily accomplished by fiber optical means.

The technologies usually involved have GaAs wafers up to 3 inch in diameter, although 5 inch diameter wafers have been made on an experimental basis. In a few cases cassette loading is being developed for 3 inch wafers, with a throughput of more than 100 wafers per week. The use of direct ion implantation into undoped semi-insulating substrates, followed by capless (or capped) annealing has been taken up in several laboratories. Vapor phase epitaxy, using arsenic trichloride, is also used in several laboratories. Future state of the art devices will require heterojunctions, so most laboratories doing compound semiconductor integrated circuit development are using molecular beam epitaxy (MBE) for research, and plan organometallic vapor phase epitaxy (OMVPE) for production. Both heterojunction modulation doped FET's (also called selectively doped FET, HEMT and TEGFET) and heterojunction bipolar transistors are being developed for logic, with most laboratories emphasizing the former.

The physical electronics of high speed GaAs transistors, the technologies used to develop them, the performance obtained to date, and some future performance expectations will be covered below.

PHYSICAL ELECTRONICS OF GaAs SHORT DEVICES

The mobility of electrons in GaAs at 300°K is about 4,500 cm^2/v-s for FET channels doped about $1 \times 10^{17}/cm^3$. In pure material, it is about 9,000 cm^2/v-s. These mobility values can be compared with 1500 cm^2/v-s for electrons in silicon. Saturation electron velocity for one-micron gate GaAs FET's is about 1.2×10^7cm/s for doped channels, and is 1.8×10^7cm/s for pure material. For very short silicon devices, where the electron velocity saturates, a value of $.6 \times 10^7$cm/s occurs. In all these cases the electron velocity is dominated by collisions with phonons and ions, or just phonons in the case of pure GaAs. At lower temperatures, electrons in pure GaAs are less likely to have the .036 eV energy to launch the dominant polar optical phonons. This allows electron mobility values of over 200,000 cm^2/v-s at 77°K and over 1,000,000 cm^2/v-s at 4°K. The electron saturation velocity controls FET performance, and reaches a value of about 3×10^7cm/s at either temperature. The frequency range over which the transistor yields current gain depends reciprocally on the electron transit time through the active portion of the transistor. In a typical GaAs doped-channel FET with a one-micron gate, this active length reaches 1.2 microns due to the electric field fringing. Thus with 1.2 x 10^7 cm/s average electron velocity, it has 10 picoseconds transit time. The unity current-gain frequency, $f_T \cong v_s/(2\pi L_e)$ where v_s is the saturation velocity and L_e is the effective electrical transit length. This yields 16 GHz for f_T, and for a higher v_s, this frequency is raised proportionately. Thus reaching higher average electron velocity values has a large impact on the speed or frequency limit of a transistor. High electron velocity also yields a proportionately large g_m, reflecting this physical effect of reduced transit time. For a given logic voltage swing and load capacitance, made up from the input capacitance of one or more following stages in parallel, as well as the interconnection leads, speed of switching rises linearly with g_m and in turn electron velocity. The loading interconnection leads are transmission lines connected to the capacitance of the gate of the next stage. This acts like any open transmission line, being simply a capacitive load as long as its length is under an eighth of a wavelength at the highest frequency component. For 5 picosecond switching speed, as long as the interconnecting line on GaAs is less than 1000 microns long, this simple equivalent circuit holds. This means that faster switching requires new designs with closer packing of circuits, with no long lines across large chips.

Electrons moving along the cube edge [100] direction in GaAs have an upper limit of group velocity of about 9.5 x 10^7 cm/s. If they are gradually accelerated with the appropriate applied voltage across pure materal less than .4 micron long, their average velocity is about 4 x 10^7 cm/s and would rise to nearly 4.75 x 10^7 cm/s for much shorter distances. This latter value is just half value of the upper limit, being a time average of a linearly rising group velocity. Electrons have about .18 micron mean free path if they have about .3 eV energy in pure GaAs, and have only about .08 micron mean free path at about .075 eV energy. The average angle of deviation during collisions with polar optical phonons is low (about 10-15°) for the .3 eV electron and is 30° for more for the lower energy electrons. This forward scattering allows a few collisions before the forward direction of the momentum is changed. If electrons are injected with .3 eV into a pure GaAs drift region, they can travel .5-.6 micron before they have lost much of their average velocity. Thus with low electric drift fields of 2,000-3,000 v/cm, to make up for the .036 eV lost to a phonon during each collision, average electron velocity values of 7-8 x 10^7 cm/s are possible over these drift distances. Donor ions can be placed in the drift regions, as long as they are well below 1 x 10^{17}/cm^3 density. This is the density at which low field mobility is halved, indicating that low energy electrons are colliding with ions as often as colliding with phonons. High energy electrons experience ion scattering cross sections that are substantially lower than those experienced by low energy electrons, of course.

In order to obtain higher average velocity on longer structures, and to approach the ballistic limit for electron velocity in shorter structures, it is necessary to have low ion density. It is also necessary to have high electron density for a high value of g_m. Using a heterojunction, with its built-in potential step, the electrons can be separated a small distance from the donor ions needed for space charge neutralization. Figure 1 shows the conduction band potential energy profile for two such

FIGURE 1

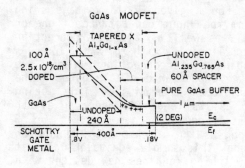

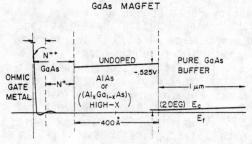

GaAs MODFET and MAGFET are represented by their conduction band potential profiles. In the MODFET the donor ions are shown as plus signs in a 100 Å Al,GaAs layer and are separated by 60 Å from the pure GaAs buffer layer. The 2 dimensional electron gas (2DEG) is formed in the pure GaAs, adjacent to the heterojunction, when positive bias is applied between the metal gate and the source of the fET. The aluminum is tapered down to zero and the top layer under the metal gate is GaAs to eliminate aluminum oxide at the interface. In the MAGFET an undoped barrier layer of AlAs (or $Al_xGa_{1-x}As$ with high X) is placed between the pure GaAs and the N^+ GaAs doped to mid $10^{17}/cm^3$. The top N^{++} GaAs layer allows an ohmic contact with an evaporated metal serving as the gate.

structures. These are designed for normally off, or enhancement mode, FET's. The first one is the MODFET or modulation doped FET. In this structure, a portion of the large band gap semiconductor Al,GaAs is doped, while the pure GaAs can have the two-dimensional electron gas (2DEG) when positive bias is applied to the metal gate. The second structure is similar to the metal-oxide-silicon FET (MOSFET) in that the metal-aluminum arsenide-gallium arsenide FET (MAGFET) layers each serve the same functions. The latter structure is less critical to grow and process with tight control of the threshold voltage. In both cases the 2DEG, of about 160 Å thickness, easily flows parallel to the heterojunction interface, due to the lack of ions in the path of the electrons. In the

MODFET, the aluminum fraction must be limited to less than .235 in the doped region to avoid the problem of the donor energy levels being deep and trapping any electrons getting into the Al,GaAs. Recently a superlattice of very thin (less than 15 Å) doped GaAs and thicker (greater than 25 Å) undoped AlAs has been used to keep the donor levels shallow compared with a conduction band, in the superlattice, but well above the conduction band housing the 2DEG.

By making very short MODFET's or MAGFET's with source-drain spacings $\leq$ .4 micron, average electron velocity values of 4 x 10^7cm/s can be reached along the channel. Thus one picosecond transit times and 160 GHz f_T values are possible. The gate length would be about .1 micron or so shorter than the source-drain spacing to avoid excess parasitic capacitance. A self-aligned process for such structures is described below.

Such very short FET's have a very substantial parasitic output conductance due to space charge limited injection current through the pure GaAs buffer layer. By thinning this buffer layer to just the thickness of the 2DEG, this output conductance can be substantially lowered. Figure 2 shows quantum well versions of the MODFET and MAGFET that would accomplish this goal. Most electrons

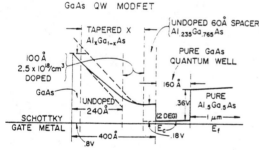

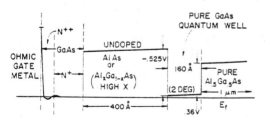

FIGURE 2

GaAs quantum well versions of both MODFET and MAGFET are represented by their conduction band potential profiles. In both cases pure $Al_{.5}Ga_{.5}As$ buffer layers are used to lower the short FET output conductance. Not only is there a .36 V barrier between the pure GaAs in the quantum well and this buffer layer, but electron saturation velocity in the $Al_xGa_{1-x}As$ buffer layer is low for $X \gtrsim .40$ further lowering the parasitic space charge limited current through the buffer layer. The quantum well thickness shown is large enough to hold the 2DEG, but could be made thicker if desired, as long as it is very thin compared to the FET source-drain spacing.

are confined to the 2DEG quantum well channel, but the few going over the barrier into the Al,GaAs buffer layer will also have substantially lower velocity. This in turn lowers any contribution to parasitic output contuctance.

Two other structures can be epitaxially grown to yield ballistic launching and drifting of electrons. One is the vertical FET and the other is the heterojunction bipolar transistor. In the FET an N^+-N heterojunction, with the doping higher in the Al,GaAs, is used to launch the electrons into the GaAs drift region. The drift region has one sixteenth as much doping, or less, as that in the Al,GaAs. The doping in the drift space is in the low to mid $10^{16}/cm^3$ for space charge neutrality with the ballistic electrons. The drift space is up to .5-.6 microns long. With the heterojunction on top of the drift region, closely spaced grooves are ion-milled and etched through this heterojunction and part way through the drift region. Using overhanging metal contacts on top, angle evaporation of the Schottky gate metal is accomplished with about .15 micron separation of the gate and the heterojunction launcher. With about $8 \times 10^7 cm/s$ average drift velocity, and .4 micron drift length, half picosecond transit time is expected. Such a device would theoretically have up to 320 GHz f_T value.

The bipolar transistor with a heterojunction emitter-base junction also has potential for high frequency operation. The Al,GaAs N type emitter, doped to $2-4 \times 10^{17}/cm^3$, can have up to .3 eV larger band gap without having deep donors. Up to .18 V of this larger band gap can be used as launching energy for ballistic electrons entering the P^+ base region. The base can have $5 \times 10^{18}/cm^3$ acceptors and be .2 microns thick, yielding low sheet resistance to aid the lowering of the base resistance. The remaining .12 V or more of heterojunction band gap difference can be used as a barrier to lower hole diffusion from the base into the emitter. Thus current gains can be high even though the base is more heavily doped than the emitter. Values of current gain from 50-200 are obtained when ballistic electron launching is used, for example. For the doping and thickness of the base shown above, the average electron velocity of $4 \times 10^7 cm/s$ is expected, yielding .5 picosecond base transit time. This is more than an order of magnitude lower than the diffusion transit time. At a design current density of $10^5 A/cm^2$, the resistance/capacitance time constant at the emitter can theoretically be as low as .3 picoseconds. For an average drift velocity of $4 \times 10^7 cm/s$, in a collector .4 microns thick, half the collector transit time is .5 psec. Thus a theoretical limit of 1.3 picosecond delay time, or an f_T value of 120 GHz is theoretically possible.

In both of these devices there needs to be extensive work on the ohmic contacts, in order to lower the parasitic resistance values limiting f_{max} and the useful switching speed for logic.

TECHNOLOGY

Gallium arsenide semi-insulating substrates can be directly ion implanted and annealed to yield reproducible, uniform FET channels. Submicron channel and gate lengths can be fabricated using self-aligned, ion-implanted ohmic contacts. Below .75 microns source-drain spacing the output conductance due to parasitic space charge limited current below the active channel, is the limiting parameter. This conductance rises as the square of the reciprocal of the source-drain spacing. The fastest such ion-implanted structure with useful logic voltage swing had 15 picoseconds switching speed at room temperature for .75 micron source-drain spacing and .55 micron gate length.[1,2] Using the same technology, a .5 micron souce-drain spacing with .3 micron gate yielded 16.7 picosecond switching speed. This latter device was limited by the output conductance. The output conductance can be lowered by placing a barrier below the active channel. Such a barrier can be a heterojunction epitaxially grown or it can be a thin layer of acceptors implanted or epitaxially grown.

The remaining devices including the MODFET, the MAGFET, as well as the ballistic injection vertical FET and heterojunction bipolar, all require epitaxial growth. Molecular beam epitaxy (MBE) has been used in the initial research, although metal organic vapor phase epitaxy (OMVPE) will be used for production. Both epitaxial growth techniques have high purity, yielding state of the art 77°K electron mobility of 125,000 to 150,000 cm^2/v-s for GaAs. Both epitaxial techniques also yield heterojunctions with 5 Å abruptness, as determined by state of the art quantum wells less than 15 Å thick showing proper photoluminescence. The quality of the Al,GaAs is also good for both techniques, now that the O_2 and H_2O can be reproducibly purged from the gases in OMVPE.[3]

In the near future, self-aligned implantation of ohmic contacts to MODFETS and MAGFETS will be optimized for high performance. Figure 3 shows the T-shaped mask technology[1] used for such a self-aligned system properly separating the contacts from the gate. The buffer layer can be made of Al,GaAs also, as shown earlier for the quantum well MODFET and MAGFET.

When there is an electric field in the top barrier during high temperature annealing, it can cause a drift in the position of the donor ions toward and even into the 2 dimensional electron gas. With a proper control of the electric field and the spacing between the ions and the 2 dimensional electron gas, the effect on the electron transport can be minimized.[4]

If the output conductance of these self-aligned, ion-implanted contact, MODFET and MAGFET devices is controlled, their switching times can be lowered to 3 picoseconds with a fan out of unity. In such short devices, the electron in the 2 dimensional electron gas will be accelerated gradually, in ballistic motion, with 4 x 10^7 cm/s average electron elocity, independent of temperature.

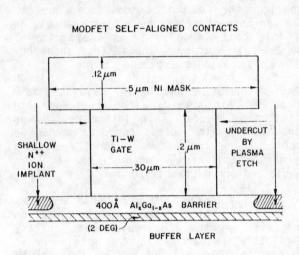

FIGURE 3
A cross section of the T-shaped gate structure for self-aligned, ion-implanted contacts is shown. Any of the MODFET or MAGFET structures, with or without the Al,GaAs buffer layers, can be processed this way. During the plasma etching of the Ti-W metal the Ni mask does not get etched. The ion implantation is shallow to allow optimum alloyed contacts to the 2DEG. Re-aligned metal deposition for alloyed contacts can be used. For a device as short as this, the buffer layer should be $Al_{.5}Ga_{.5}As$ to prevent high output conductance.

CONCLUSIONS

Compound semiconductors like GaAs have high average velocity at relatively low voltage values when collisions dominate transport, but with sub-half-micron dimensions there can be near ballistic electron transport with relatively low voltage. The high velocity values raise g_m values and frequency response for given values of parasitic capacitance. For properly shortened interconnection lines, logic switching times below 10 picoseconds, and even below 5 picoseconds, are eventually expected for unity fan out. These switching times will be raised by increased fan out, but will always be lower than the case for silicon, in proportion to the switching voltage and the reciprocal of the average electron velocity. For the present, self-aligned ion-implanted GaAs MESFETS will perform well, even down to .5 micron source-drain spacing. In the near future, MODFETS and MAGFETS using self-aligned, ion-implanted contacts will have from 1.5 to 3 times as high electron velocity at room temperature, dominating the high speed logic I.C. In the long term, vertical FET and heterojunction bipolars will have further impact on high speed I.C.'s. Epitaxial growth of heterojunctions for I.C.'s will be necessary, and OMVPE will eventually be needed for high volume production. Three inch diameter substrates will be normal for I.C.'s in the near future, using cassettes for ease of handling.

REFERENCES

1. R.A. Sadler and L.F. Eastman, Electron Device Lett., EDL-4 (7) 215-217 (July 1983).

2. R. Sadler, Ph.D. Thesis, Cornell University, Ithaca, NY (Jan. 1984).

3. J.R. Shealy, Ph.D. Thesis, Cornell University, Ithaca, NY (May 1983).

4. H. Lee, G. Wicks and L.F. Eastman, Proc. Ninth Biennial High Speed Semiconductor Devices and Circuits Conf., Cornell University, Ithaca, NY (Aug. 15-17, 1983).

ACKNOWLEDGEMENTS

This work was supported by the Air Force Office of Scientific Research, DARPA and IBM.

ELECTRON SCATTERING AND MOBILITY IN A QUANTUM WELL HETEROLAYER

Vijay K. Arora* and Athar Naeem
Department of Physics, King Saud University
P.O. Box 2455, Riyadh, Saudi Arabia

ABSTRACT

The theory of electron-lattice scattering is analyzed for a quantum-well heterolayer under the conditions that the de Broglie wavelength of an electron is comparable to or larger than the width of the layer, and donor impurities are removed in an adjacent nonconducting layer. The mobility due to isotropic scattering by acoustic phonons, point defects, and alloy scattering is found to increase whereas that due to polar-optic phonon scattering is found to decrease with increasing thickness.

INTRODUCTION

These days, there is an intense interest in the design and development of new high-speed devices used in VLSI circuits. This is mainly due to the increased cleanliness associated with the new technologies of Molecular Beam Epitaxy (MBE), Metalorganic Chemical Vapor Deposition (MOCVD) and the fine line lithography. An experimental GaAs-AlGaAs high electron mobility transistor (HEMT) has recently been realized using these growth and characterization techniques.[1] The switching time for this HEMT is a little more than 10 picoseconds and it generates little heat, making it an ideal component for a supercomputer. One of the reasons for high mobility in a quantum-well heterolayer of which HEMT is made of is the removal of the donor ionized impurities in an adjacent nonconducting heterolayer with electron scattering considerably reduced[2,3] especially at low temperatures. Stormer[4] has listed the possible scattering interactions which are likely to play a role in limiting the mobility in a heterolayer. With ionized impurity scattering considerably reduced, the mobility is limited by electron scattering via unavoidable defects, alloy structure, and lattice phonons, which are examined in this paper.

ISOTROPIC SCATTERING

The scattering due to the acoustic phonon scattering[5], neutral point defects,[5,6] and that due to potential fluctuations in the alloy structure[6] is isotropic in the sense that the electrons have no preferential direction on scattering. The electron scattering by acoustic phonons is one of the important isotropic mechanisms of

*Visitor at GTE Laboratories, 40 Sylvan Road, Waltham, MA 02254

0094-243X/84/1220280-06 $3.00 Copyright 1984 American Institute of Phy scattering. Price[7] has considered the piezoelectric coupling as well as the deformation potential coupling. Basu and Nag[8] have shown that the piezoelectric scattering by acoustic phonons does not play an active role at intermediate temperatures. The most accepted expression for the mobility limited by the deformation potential scattering via acoustic phonons in a quantum-well heterolayer is given by[2,5]

$$\mu_a = 2\,e\,\hbar^3\,\rho_d u^2 d/3\,m^{*2}\,E_1\,k_B T, \tag{1}$$

where ρ_d is the crystal density, u is the longitudinal sound velocity, d is the width of the quantum well heterolayer considered to be smaller than the de Broglie wavelength of an electron, m^* is the effective mass, E_1 is the deformation potential constant, and T is the temperature. There is a certain degree of uncertainty concerning the value of E_1. With the value of E_1 = 7 eV, quoted by Price,[7] μ_a is calculated as

$$\mu_a = 2.9 \times 10^6\,\frac{d(nm)}{T(K)}\,\frac{cm^2}{V\text{-}s}. \tag{2}$$

With typical values of d = 10 nm and T = 100 K, μ_a = 2.9 x 10^5 cm^2/V-s.

At low temperatures, the point-defect scattering and alloy scattering may become important in limiting the mobility. Simplified models[6] of both of these scattering mechanisms give for mobilities the expressions:

$$\mu_{pd} = 2\,e\,\hbar^3 d/3\,m^{*2} N_d V_o^2, \tag{3}$$

$$\mu_{al} = 2\,e\,\hbar^3 N_{al}\,d/3\,m^{*2}\,\alpha(1-\alpha)\Delta E^2, \tag{4}$$

where N_d is the number of point defects, and V_o is the potential parameter associated with a defect, N_{al} is the volume concentration of atoms in the alloy, α is the relative concentration of one kind of atoms in a binary alloy, and ΔE is the conduction band discontinuity at the edge of a heterolayer.

As Stormer[4] has pointed out that MBE technique is being perfected to the point that the number of point defects can be considerably reduced, so that the mobility contribution due to Eq.(3) can easily be neglected. But alloy scattering is still present as the quantum well heterojunction has an alloy composition on a side. Considering $\Delta E = 300$ meV and $\alpha = 0.3$ gives a mobility value $\mu_{al} = 1.5 \times 10^5$ cm^2/V-s for d = 10 nm. Therefore, the alloy scattering is comparable to the acoustic-phonons scattering at intermediate temperature and dominates it at low temperature.

POLAR OPTICAL PHONON SCATTERING

In a quantum-well heterostructure made of polar materials, the polar-optical-phonon (POP) scattering is considered important.[4,7,9] One of the problems with the theory of the optical phonon scattering is the difficulty in defining the momentum relaxation rate in the Boltzmann transport framework when the scattering of carriers is highly inelastic. Due to this difficulty, a relaxation rate is normally calculated. The relaxation rate τ_P^{-1} calculated by Price for sufficiently thin layers is given by the expressions:

$$\tau_P^{-1} = (\pi e^2 k_0 / 2 \varepsilon_p \hbar) (N_0 + \tfrac{1}{2} \pm \tfrac{1}{2})^{\tfrac{1}{2}}, \tag{5}$$

with $k_0 = (2 m^* \omega_0 / \hbar)^{\tfrac{1}{2}}$, (6)

$$\varepsilon_p^{-1} = (\varepsilon_\infty^{-1} - \varepsilon_s^{-1}), \tag{7}$$

Here N_0 is the temperature-dependent number of optical phonons, $\hbar \omega_0$ is the energy of an optical phonon, ε_∞ is the high frequency dielectric constant, and ε_s is the static dielectric constant. The upper negative sign is for absorption, and the lower positive sign is for emission of an optical phonon. It may be noted that emission scattering rate of Eq.(5) is zero if the carrier kinetic energy in a heterolayer $\varepsilon_k > \hbar \omega_0$, so that the phonon emission is not possible. The mobility due to this scattering can then be calculated in a straightforward manner with the resulting expression:

$$\mu_{POP}^P = \mu_A^P [1 + \exp(-x_0)(1 + x_0)(N_0/(2 N_0 + 1)) - 1], \tag{8}$$

with

$$\mu_A^P = 2 \hbar \varepsilon_p / m^* k_0 e \pi N_0 \tag{9}$$

$$x_0 = \hbar \omega_0 / k_B T \tag{10}$$

where μ_A is the mobility due to phonon absorption only, whose value at T = 300 K for parameters[7] appropriate to GaAs is $\mu_A^P = 0.07 \times 10^5$ cm^2/V-s and that at T = 100 K is $\mu_A^P = 1.4 \times 10^5$ cm^2/V-s. This means if scattering rate derived by Price are any indications of limiting the mobility, POP scattering is dominant at room temperature and may become comparable to acoustic phonons scattering at intermediate temperatures.

Ridley[9] has indicated an important distinction between scattering rate and momemtum relaxation rate, which seems to be relevant to an anisotropic term in the density matrix formalism.[10] This makes momentum relaxation rate distinct from scattering rate. If this distinction is considered, the momemtum relaxation rate (after correction of a factor of two and using CGS units) is given by:

$$\tau_P^{-1} = e^2 \omega_o N_o m^* d/4 \pi \varepsilon_p \hbar^2 \text{ (absorption)}, \tag{11}$$

$$= \pi e^2 \omega_o (N_o + 1)/4 \varepsilon_p d \varepsilon_k \text{ (emission)}. \tag{12}$$

When these are considered in the calculation of mobility, the resulting expression is:

$$\mu = \mu_A [1-\delta \exp(-x_o)-\delta^2 \exp(\delta) E_1(x_o + \delta)], \tag{13}$$

with $\mu_A^R = 4 \pi \varepsilon_p \hbar^2/e \omega_o N_o m^{*2} d,$ (14)

$$\delta = 8(N_o + 1)\varepsilon_o/N_o k_B T, \tag{15}$$

where μ_A^R is the mobility expression due to optical phonon absorption only whose value at 300 K is $\mu_A^R = 1.0 \times 10^5$ cm^2/V-s, and that at 100 K is $\mu_A^R = 2.2 \times 10^6$ cm^2/V-s. Therefore, in Ridley's model the POP scattering is negligible at intermediate temperatures, but becomes comparable to acoustic phonon scattering at room temperature. $E_1(x)$ in Eq.(13) is the exponential integral of order 1, and ε_o is the zero point energy in the quantum well.

DISCUSSION AND SUMMARY

The results presented here are valid under conditions of size-quantum limit when most of the electrons occupy the lowest quantized subband. The separation between the two lowest subbands is 168 meV for d = 10 nm, which is much larger than the thermal energy at room temperature.

We show in Fig. 1, a log-plot of mobility limited by acoustic phonons, by absorption of polar optical phonons (Eq.(14)), and the combined mobility of acoustic phonons, optical phonons, and temperature-independent alloy scattering ($\mu_{al} = 1.6 \times 10^5$ cm^2/V-s). It is clear from the graph that the temperature-dependent alloy scattering plays a dominant role at low temperatures, acoustic phonons at intermediate temperatures, and POP scattering may become important at room temperatures. At lower temperatures, phonon emission is negligible because the number of carriers able to emit optical phonons is relatively small. But at higher temperatures, there may be an appreciable contribution due to electron scattering by emission of an optical phonon as then the rate of emission and absorption tends to be equal.

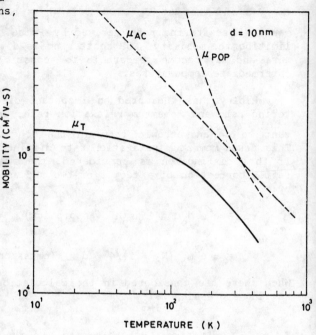

Fig. 1 the mobility in a quantum-well limited by scattering due to acoustic phonons, polar optical phonons (absorption), and combined mobility including alloy scattering.

The results presented above are valid when there is no transverse gate field present. In the presence of a high electric field, the carrier distribution is distinctly different from the equilibrium distribution function.[11] Then, the longitudinal current may be affected by the presence of a dipole layer across the heterojunction.

In a device design, the critical thickness for maximum mobility may be an important consideration. The mobility limited by isotropic scattering decreases, whereas the mobility limited by POP scattering (Eq.(13)) increases with the decreasing thickness. When both isotropic scattering and POP scattering are considered, an optimal thickness for maximum mobility is obtained.

ACKNOWLEDGEMENT

One of us (V.K.A.) is thankful to Drs. Scott Norman and Johnson Lee for access to their facilities during his visit to GTE Laboratories.

REFERENCES

1. H. Morkoc and P. M. Solomon, IEEE Spectrum 21 (2), 28 (February 1984).
2. J. Lee, H. N. Spector, and V. K. Arora, J. Appl. Phys. 54, 6995 (1983).
3. H. N. Spector and V. K. Arora, unpublished.
4. H. L. Stormer, Surf. Sci. 132, 519 (1983).
5. V. K. Arora and F. G. Awad, Phys. Rev. B 23, 5570 (1981).
6. V. K. Arora and H. N. Spector, J. Appl. Phys. 54, 831 (1983).
7. P. J. Price, Ann. Phys. 133, 217 (1981).
8. P. K. Basu and B. R. Nag, J. Phys. C: Solid State Phys. 14, 1519 (1981).

CHARACTERIZATION OF DEEP LEVEL DEFECTS IN REACTIVE ION BEAM ETCHED InP

Y. Yuba, K. Gamo, Y. Judai and S. Namba
Faculty of Engineering Science, Osaka University
Toyonaka, Osaka 560, Japan

ABSTRACT

Ion beam etching(IBE) of InP was carried out by using 0.5 to 2 keV Ar or Cl_2 ions. A deep level center associated with IBE induced defect with an activation energy of 0.19 eV as well as those involved in the starting materials were resolved by an optical transient current spectroscopy(OTCS) in the sample after Ar IBE. Sheet resistance of semi-insulating(SI) sample showed a significant decrease after IBE using both Ar and Cl_2 gases and this was attributed to the introduction of the 0.19 eV center over an Fe-related compensation center. The sheet resistance was observed to recover after annealing at temperatures around 300 C. From conductivity profile measurements, it was found that Cl_2 reactive IBE induced defects with a smaller concentration and a shallower profile than Ar IBE at the same energy and ion dose due to the difference in etching rate.

INTRODUCTION

Various dry etchings including ion beam etching(IBE) have become of great interest as the minimum linewidth required in device fabrication is reduced to 1 μm or less. The use of an ion beam rather than the discharge as in plasma etching offers several potential advantages: Anisotropic processing and excellent pattern transfer are readily achievable, much better control over ion energy, current density and the angle of incidence is available and ion energy and current density can be monitored during operation and set independently over a reasonable range. These features yield unique properties of IBE suitable for microfabrication techniques. In Ⅲ-V compound semiconductors, IBE has been expected to be a promising technique for fabrication of lasers without a conventional cleaved cavity-mirror and their monolithic integration.[1] We have previously shown from a series of studies[2,3] on IBE of InP that the Cl_2 reactive IBE is superior in etching characteristics. Also we have fabricated by IBE grating structures with a period around 2500 A essential for the integrated optics devices to demonstrate its feasiblity as a microfabrication technique.

One drawback of IBE is that the ion irradiation on the surface of the target material, even if at low energy, inherently induces damages, which may affect various properties of the material. Therefore, in this report, we will describe the spectroscopic characterization and comparison of deep levels associated with damage in InP induced by Ar and Cl_2 reactive IBE.

EXPERIMENTALS

Two type (100) oriented wafers of LEC-grown InP single-crystals were used in the present experiment; one was Fe-doped semi-insulating(SI) wafer with a resistivity above 10^6 ohm-cm at room temperature in dark and the other was Si- or S-doped n-type wafer with a carrier concentration between 5×10^{15} and 1×10^{17} /cm^3. They were mechanically polished and subsequently etched in a Br-methanol solution before IBE.

Ion beam etching was performed by use of an ion shower apparatus[4] with a Kaufman type ion source, which yielded an ion beam with a diameter around 3 cm. Samples were set on a water-cooled stage. Ion energy ranged from 0.5 to 2.0 keV and beam current density from 0.2 to 0.28 mA/cm^2. Source gases used were commercially available high purity Ar and Cl_2 and they were introduced into the ion source through a feed-back controlled automatic needle valve. Vapor pressure in the target chamber was maintained 10^{-4} mmHg during etching. Isochronal annealing was performed in flowing pure hydrogen atmosphere up to 700 C.

Deep level characterization was performed by an optical transient current spectroscopy(OTCS)[5] on the sample formed on SI substrate or deep level transient spectroscopy(DLTS)[6] on MOS structure formed on the n-type substrate by anodic oxidation technique. Sheet resistance and conductivity profile were measured on the sample formed using SI InP. Details of measurements were described elsewhere.[7]

RESULTS AND DISCUSSION

Figure 1 shows typical OTCS spectra of the samples before and after Ar and Cl_2 IBE at 1 keV and 0.28 mA/cm^2 for 80 sec. After Ar IBE, a new center L-3 was clearly resolved in addition to the center L-1, which was related with a dominant dopant Fe. From the temperature dependence of thermal emission rate, we obtained an activation energy for each detected center as shown in Fig. 1. The L-3 center has an activation energy of 0.19 eV, which shows a fairly good agreement with that of

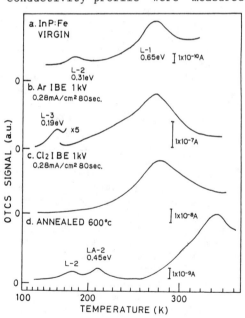

Fig.1. OTCS spectra of as-received, ion beam etched and annealed SI-InP.

the defect center induced by the 60 keV ion irradiation [7]. We obtained from DLTS measurements, using n-type sample evidence that this center had the character of electron trap. These results suggested that the same defect center as that observed in high energy ion irradiated sample was introduced by IBE, irrespective of difference in ion energy and ion dose. In OTCS spectrum of the sample after the Cl_2 IBE at the same condition, we detected only the centers involved in the starting material and no evidence of the 0.19 eV defect center. This proposed damage effect was small in case of Cl_2 IBE.

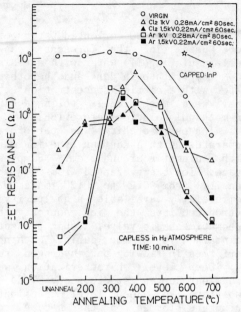

Fig.2. Annealing behavior of sheet resistance for as-received and ion beam etched SI samples.

Figure 2 shows the annealing behavior of the sheet resistance for the SI sample after Ar and Cl_2 IBE. The result of the untreated substrate is also presented for comparison. The sheet resistance showed a significant decrease after IBE. This was more pronounced in case of Ar IBE and also both samples etched by Ar and Cl_2 IBE at higher energy showed the larger decrease.

We have found previously in 60 keV-ion-irradiated, Fe-doped, SI InP the introduction of the 0.19 eV defect center results in the decrease of sheet resistance because this center has a shallower level than that of Fe center responsible for high resistivity of SI InP and therefore can yield free carriers if it dominates over the deep compensation center.[7] This mechanism was supported by a good agreement of activation energy for carrier emission from the center and for sheet resistance as well as a strong correlation observed in fluence dependence and annealing behavior. This seems to be the case also for IBE and it is natural to conclude the decrease of sheet resistance is caused by the introduction of the 0.19 eV defect center. In the sample etched by Cl_2 IBE, even though the 0.19 eV center was not resolved clearly by OTCS, the result of sheet resistance measurement suggested its presence.

Sheet resistance showed a recovery after the annealing up to temperatures of 300 to 400 C and attained the value about one order of magnitude smaller than that of the starting material. A similar annealing behavior, consistent with that for the 0.19 eV center observed in 60 keV ion irradiated samples, was obtained in both cases of Ar and Cl_2 IBE and this suggested the same defect center was present. This annealing behavior also seemed to correspond well with that for a crystalline quality described previously [2]. The annealing above 400 C again gave rise to a decrease of sheet resistance and a similar tendency, though small, was observed in the untreated substrate after the annealing. Therefore this decrease was considered to be due to the annealing induced effect.

Figure 3 shows depth distributions of conductivity in the samples after 1 keV Ar and Cl_2 IBE for 3 min. These data were obtained by successive sheet resistance measurements combined with a layer removal technique using

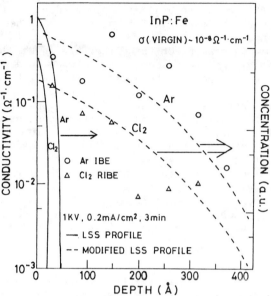

Fig.3. Depth distributions of conductivity, LSS theoretical profiles of retained ions(solid lines) and modified profiles incluiding enhanced diffusion (dashed lines). Details see the text.

anodic oxidation. Solid lines represent theoretical depth profiles of retained ions estimated by LSS theory[8] in taking into account the erosion of surface with a measured etching rate[3] , where the parameters used for LSS Gaussian distribution[8] are the projected ranges of 19.4 and 12.5 A, and the standard deviations of 13.6 and 9.3 A, respectively, for Ar and Cl_2 ions. Dashed lines show the profiles of ions obtained in the same manner as before except adding an extra 150 A to the standard deviations for taking a possible diffusion in consideration. As discussed above, the increase of conductivity is due to the IBE induced 0.19 eV center and therefore conductivity profile can be considered to represent a depth distribution of induced defect simply if mobility has no appreciable depth dependence. It is clear the conductivity increased in the region deeper than the LSS projected range[8]. We observed that the difference in sheet resistance still remained at a depth of 600 A or more. The data showed a sufficient agreement, regardless of a simple model and various unknown factors, with the modified theoretical profiles including broadening due to diffusion rather than the LSS profiles. These results suggested the diffusion of defects occured during IBE, which was small compared with that of GaAs[2]. Furthermore, it was noticed that the conductivity increased more significantly at a given depth for Ar IBE than Cl_2 IBE. Therefore, we can conclude that the Ar IBE induces defects with a higher concentration over a deeper region than Cl_2 IBE at the same condition on energy, current density and etching time. If we apply IBE for the etching of a given depth, a resultant damage effect should be more significant for Ar IBE because of its low etching rate.

A reason for the difference in the defect profile observed is considered to be due to difference in etching rate. As shown in Fig. 3, ions which remain after Cl_2 IBE have a smaller concentration and a shallower profile than those after Ar IBE at the same IBE condition since a larger part of ions locate in the region etched off finally. If we assume the amount of defects is proportional to ion dose, the same situation seems to exist also for the induced defect, though defect profile is broader because of an enhanced diffusion. It is expected to be difficult to reduce IBE induced defects completely since ion irradiation is not avoidable. However, IBE induced defects in InP can be removed readily by annealing as shown in Fig. 2 also we consider another practical approach for etching free of induced defect is the combination of IBE with a subsequent wet chemical or anodic etching.

CONCLUSIONS

We have investigated deep levels associated with damage induced in InP by IBE by means of OTCS, DLTS and sheet resistance measurements.

A deep level acting as an electron trap with an activation energy of 0.19 eV is observed to be induced by IBE and to show an annealing stage at temperatures around 300 C. From the profile measurement of this deep level, it is found that the induced defect has a higher concentration and a deeper profile in case of Ar IBE compared with Cl_2 IBE at the same IBE condition. Therefore it can be concluded that Cl_2 IBE is superior for induced defect to Ar IBE.

Finally, the authors wish to thank M. Takai for helpful discussions, H. Aritome for the use of the apparatus, and K. Mino and K. Matsubara for technical assistance.

REFERENCES

1. For example, K. Asakawa and S. Sugata, Jpn. J. Appl. Phys. 22, L653 and references cited therein.
2. Y. Yuba, K. Gamo, H. Toba, Xi Guan He and S. Namba, Jpn. J. Appl. Phys. 22, 1206 (1983).
3. Y. Yuba, K. Gamo, Xi Guan He, Yu Shu Zhang and S. Namba, Jpn. J. Appl. Phys. 22, 1211 (1983).
4. S. Matsui, T. Yamato, H. Aritome and S. Namba, Jpn J. Appl. Phys. 19, L126 (1980).
5. Ch. Hurtes, M. Boulou, A. Mitonneau and D. Bois, Appl. Phys. Lett. 32, 821 (1978).
6. D. V. Lang, J. Appl. Phys. 45, 3023 (1974).
7. Y. Yuba, Y. Judai, K. Gamo and S. Namba, Physica, 116B, 461 (1983).
8. K. B. Winterbon, Ion Implantation Range and Energy Deposition Distribution (IFI/Plenum, N. Y., 1975).

AUTHOR INDEX

Arora, V.K. - 186,280

Bean, J.C. - 198

Carlos, W.E. - 34
Cerofolini, G.F. - 225
Climent, A. - 63
Coburn, J.W. - 98
Craven, R.A. - 205

Dix, C. - 92
Dorda, G. - 167

Eastman, L.F. - 271

Ferry. D.K. - 162,220
Flavin, P.G. - 92
Fonash, S.J. - 63,106

Gamo, K. - 286
Garrigues, M. - 39
Gibbons, J.F. - 20
Gondin, R.O. - 220

Habermeier, H.-U. - 192
Hattori, T. - 45
He, D. - 45
Hendy, P. - 92
Hughes, H.L. - 34

Jones, M.E. - 92
Judai, Y. - 286

Kawazu, S. - 120
Koch, F. - 50
Koyama, H. - 120
Krawczyk, S. - 39

Larrabee, G.B. - 139
Linsker, R. - 251
Lugli, P. - 162
Luo, G. - 145
Lyon, S.A. - 8

Mazur, J.T. - 52
McGill, T.C. -181
Mrabeut, T. - 39
Mu, X.C. - 63
Muto, M. - 45

Naeem, A. - 280
Namba, S. - 286
Ng, J. - 20
Ning, T.H. - 151
Nishioka, T. - 120
Noble, W.P. - 156

Pantelides, S.T. - 125
Polignano, M.L. - 225
Porad, W. - 220
Prabhakar, A. - 181

Ravaiolo, U. - 162
Rigaud, D. - 240

Schmidt-Landsiedel, D. - 167
Shao, H. - 145
Shi, Z. - 145
Shimura, F. - 205
Sigmon, T. - 20
Sodini, D. - 240
Solomon, P.M. - 172
Suzuki, T. - 45

Touboul, A. - 240

von Gutfeld, R.J. - 56

Walker, W.W. - 156
Wang, W. - 145
Washburn, J. - 52
White, S.R. - 261
Wilson, A.D. - 69
Winters, H.F. - 98

Xu, K. - 145

AUTHOR INDEX

Yamabe, K. - 45
Yamauchi, H. - 45
Yi, K.S. - 220
Young, D.R. - 1
Yuba, Y. - 286

Zrenner, A. - 50

AIP Conference Proceedings

		L.C. Number	ISBN
No. 1	Feedback and Dynamic Control of Plasmas	70-141596	0-88318-100-2
No. 2	Particles and Fields - 1971 (Rochester)	71-184662	0-88318-101-0
No. 3	Thermal Expansion - 1971 (Corning)	72-76970	0-88318-102-9
No. 4	Superconductivity in d-and f-Band Metals (Rochester, 1971)	74-18879	0-88318-103-7
No. 5	Magnetism and Magnetic Materials - 1971 (2 parts) (Chicago)	59-2468	0-88318-104-5
No. 6	Particle Physics (Irvine, 1971)	72-81239	0-88318-105-3
No. 7	Exploring the History of Nuclear Physics	72-81883	0-88318-106-1
No. 8	Experimental Meson Spectroscopy - 1972	72-88226	0-88318-107-X
No. 9	Cyclotrons - 1972 (Vancouver)	72-92798	0-88318-108-8
No. 10	Magnetism and Magnetic Materials - 1972	72-623469	0-88318-109-6
No. 11	Transport Phenomena - 1973 (Brown University Conference)	73-80682	0-88318-110-X
No. 12	Experiments on High Energy Particle Collisions - 1973 (Vanderbilt Conference)	73-81705	0-88318-111-8
No. 13	π-π Scattering - 1973 (Tallahassee Conference)	73-81704	0-88318-112-6
No. 14	Particles and Fields - 1973 (APS/DPF Berkeley)	73-91923	0-88318-113-4
No. 15	High Energy Collisions - 1973 (Stony Brook)	73-92324	0-88318-114-2
No. 16	Causality and Physical Theories (Wayne State University, 1973)	73-93420	0-88318-115-0
No. 17	Thermal Expansion - 1973 (lake of the Ozarks)	73-94415	0-88318-116-9
No. 18	Magnetism and Magnetic Materials - 1973 (2 parts) (Boston)	59-2468	0-88318-117-7
No. 19	Physics and the Energy Problem - 1974 (APS Chicago)	73-94416	0-88318-118-5
No. 20	Tetrahedrally Bonded Amorphous Semiconductors (Yorktown Heights, 1974)	74-80145	0-88318-119-3
No. 21	Experimental Meson Spectroscopy - 1974 (Boston)	74-82628	0-88318-120-7
No. 22	Neutrinos - 1974 (Philadelphia)	74-82413	0-88318-121-5
No. 23	Particles and Fields - 1974 (APS/DPF Williamsburg)	74-27575	0-88318-122-3
No. 24	Magnetism and Magnetic Materials - 1974 (20th Annual Conference, San Francisco)	75-2647	0-88318-123-1
No. 25	Efficient Use of Energy (The APS Studies on the Technical Aspects of the More Efficient Use of Energy)	75-18227	0-88318-124-X

AIP Conference Proceedings

No.	Title	LCCN	ISBN
No. 26	High-Energy Physics and Nuclear Structure - 1975 (Santa Fe and Los Alamos)	75-26411	0-88318-125-8
No. 27	Topics in Statistical Mechanics and Biophysics: A Memorial to Julius L. Jackson (Wayne State University, 1975)	75-36309	0-88318-126-6
No. 28	Physics and Our World: A Symposium in Honor of Victor F. Weisskopf (M.I.T., 1974)	76-7207	0-88318-127-4
No. 29	Magnetism and Magnetic Materials - 1975 (21st Annual Conference, Philadelphia)	76-10931	0-88318-128-2
No. 30	Particle Searches and Discoveries - 1976 (Vanderbilt Conference)	76-19949	0-88318-129-0
No. 31	Structure and Excitations of Amorphous Solids (Williamsburg, VA., 1976)	76-22279	0-88318-130-4
No. 32	Materials Technology - 1976 (APS New York Meeting)	76-27967	0-88318-131-2
No. 33	Meson-Nuclear Physics - 1976 (Carnegie-Mellon Conference)	76-26811	0-88318-132-0
No. 34	Magnetism and Magnetic Materials - 1976 (Joint MMM-Intermag Conference, Pittsburgh)	76-47106	0-88318-133-9
No. 35	High Energy Physics with Polarized Beams and Targets (Argonne, 1976)	76-50181	0-88318-134-7
No. 36	Momentum Wave Functions - 1976 (Indiana University)	77-82145	0-88318-135-5
No. 37	Weak Interaction Physics - 1977 (Indiana University)	77-83344	0-88318-136-3
No. 38	Workshop on New Directions in Mossbauer Spectroscopy (Argonne, 1977)	77-90635	0-88318-137-1
No. 39	Physics Careers, Employment and Education (Penn State, 1977)	77-94053	0-88318-138-X
No. 40	Electrical Transport and Optical Properties of Inhomogeneous Media (Ohio State University, 1977)	78-54319	0-88318-139-8
No. 41	Nucleon-Nucleon Interactions - 1977 (Vancouver)	78-54249	0-88318-140-1
No. 42	Higher Energy Polarized Proton Beams (Ann Arbor, 1977)	78-55682	0-88318-141-X
No. 43	Particles and Fields - 1977 (APS/DPF, Argonne)	78-55683	0-88318-142-8
No. 44	Future Trends in Superconductive Electronics (Charlottesville, 1978)	77-9240	0-88318-143-6
No. 45	New Results in High Energy Physics - 1978 (Vanderbilt Conference)	78-67196	0-88318-144-4
No. 46	Topics in Nonlinear Dynamics (La Jolla Institute)	78-057870	0-88318-145-2
No. 47	Clustering Aspects of Nuclear Structure and Nuclear Reactions (Winnepeg, 1978)	78-64942	0-88318-146-0
No. 48	Current Trends in the Theory of Fields (Tallahassee, 1978)	78-72948	0-88318-147-9
No. 49	Cosmic Rays and Particle Physics - 1978 (Bartol Conference)	79-50489	0-88318-148-7

AIP Conference Proceedings

No. 50	Laser-Solid Interactions and Laser Processing - 1978 (Boston)	79-51564	0-88318-149-5
No. 51	High Energy Physics with Polarized Beams and Polarized Targets (Argonne, 1978)	79-64565	0-88318-150-9
No. 52	Long-Distance Neutrino Detection - 1978 (C.L. Cowan Memorial Symposium)	79-52078	0-88318-151-7
No. 53	Modulated Structures - 1979 (Kailua Kona, Hawaii)	79-53846	0-88318-152-5
No. 54	Meson-Nuclear Physics - 1979 (Houston)	79-53978	0-88318-153-3
No. 55	Quantum Chromodynamics (La Jolla, 1978)	79-54969	0-88318-154-1
No. 56	Particle Acceleration Mechanisms in Astrophysics (La Jolla, 1979)	79-55844	0-88318-155-X
No. 57	Nonlinear Dynamics and the Beam-Beam Interaction (Brookhaven, 1979)	79-57341	0-88318-156-8
No. 58	Inhomogeneous Superconductors - 1979 (Berkeley Springs, W.V.)	79-57620	0-88318-157-6
No. 59	Particles and Fields - 1979 (APS/DPF Montreal)	80-66631	0-88318-158-4
No. 60	History of the ZGS (Argonne, 1979)	80-67694	0-88318-159-2
No. 61	Aspects of the Kinetics and Dynamics of Surface Reactions (La Jolla Institute, 1979)	80-68004	0-88318-160-6
No. 62	High Energy e^+e^- Interactions (Vanderbilt, 1980)	80-53377	0-88318-161-4
No. 63	Supernovae Spectra (La Jolla, 1980)	80-70019	0-88318-162-2
No. 64	Laboratory EXAFS Facilities - 1980 (Univ. of Washington)	80-70579	0-88318-163-0
No. 65	Optics in Four Dimensions - 1980 (ICO, Ensenada)	80-70771	0-88318-164-9
No. 66	Physics in the Automotive Industry - 1980 (APS/AAPT Topical Conference)	80-70987	0-88318-165-7
No. 67	Experimental Meson Spectroscopy - 1980 (Sixth International Conference, Brookhaven)	80-71123	0-88318-166-5
No. 68	High Energy Physics - 1980 (XX International Conference, Madison)	81-65032	0-88318-167-3
No. 69	Polarization Phenomena in Nuclear Physics - 1980 (Fifth International Symposium, Santa Fe)	81-65107	0-88318-168-1
No. 70	Chemistry and Physics of Coal Utilization - 1980 (APS, Morgantown)	81-65106	0-88318-169-X
No. 71	Group Theory and its Applications in Physics - 1980 (Latin American School of Physics, Mexico City)	81-66132	0-88318-170-3
No. 72	Weak Interactions as a Probe of Unification (Virginia Polytechnic Institute - 1980)	81-67184	0-88318-171-1
No. 73	Tetrahedrally Bonded Amorphous Semiconductors (Carefree, Arizona, 1981)	81-67419	0-88318-172-X
No. 74	Perturbative Quantum Chromodynamics (Tallahassee, 1981)	81-70372	0-88318-173-8

AIP Conference Proceedings

No. 75	Low Energy X-ray Diagnostics-1981 (Monterey)	81-69841	0-88318-174-6
No. 76	Nonlinear Properties of Internal Waves (La Jolla Institute, 1981)	81-71062	0-88318-175-4
No. 77	Gamma Ray Transients and Related Astrophysical Phenomena (La Jolla Institute, 1981)	81-71543	0-88318-176-2
No. 78	Shock Waves in Condensed Matter - 1981 (Menlo Park)	82-70014	0-88318-177-0
No. 79	Pion Production and Absorption in Nuclei - 1981 (Indiana University Cyclotron Facility)	82-70678	0-88318-178-9
No. 80	Polarized Proton Ion Sources (Ann Arbor, 1981)	82-71025	0-88318-179-7
No. 81	Particles and Fields - 1981: Testing the Standard Model (APS/DPF, Santa Cruz)	82-71156	0-88318-180-0
No. 82	Interpretation of Climate and Photochemical Models, Ozone and Temperature Measurements (La Jolla Institute, 1981)	82-071345	0-88318-181-9
No. 83	The Galactic Center (Cal. Inst. of Tech., 1982)	82-071635	0-88318-182-7
No. 84	Physics in the Steel Industry (APS.AISI, Lehigh University, 1981)	82-072033	0-88318-183-5
No. 85	Proton-Antiproton Collider Physics - 1981 (Madison, Wisconsin)	82-072141	0-88318-184-3
No. 86	Momentum Wave Functions - 1982 (Adelaide, Australia)	82-072375	0-88318-185-1
No. 87	Physics of High Energy Particle Accelerators (Fermilab Summer School, 1981)	82-072421	0-88318-186-X
No. 88	Mathematical Methods in Hydrodynamics and Integrability in Dynamical Systems (La Jolla Institute, 1981)	82-072462	0-88318-187-8
No. 89	Neutron Scattering - 1981 (Argonne National Laboratory)	82-073094	0-88318-188-6
No. 90	Laser Techniques for Extreme Ultraviolt Spectroscopy (Boulder, 1982)	82-073205	0-88318-189-4
No. 91	Laser Acceleration of Particles (Los Alamos, 1982)	82-073361	0-88318-190-8
No. 92	The State of Particle Accelerators and High Energy Physics (Fermilab, 1981)	82-073861	0-88318-191-6
No. 93	Novel Results in Particle Physics (Vanderbilt, 1982)	82-73954	0-88318-192-
No. 94	X-Ray and Atomic Inner-Shell Physics-1982 (International Conference, U. of Oregon)	82-74075	0-88318-193
No. 95	High Energy Spin Physics - 1982 (Brookhaven National Laboratory)	83-70154	0-88318-19
No. 96	Science Underground (Los Alamos, 1982)	83-70377	0-88318-1 -9

No. 97	The Interaction Between Medium Energy Nucleons in Nuclei-1982 (Indiana University)	83-70649	0-88318-196-7
No. 98	Particles and Fields - 1982 (APS/DPF University of Maryland)	83-70807	0-88318-197-5
No. 99	Neutrino Mass and Gauge Structure of Weak Interactions (Telemark, 1982)	83-71072	0-88318-198-3
No. 100	Excimer Lasers - 1983 (OSA, Lake Tahoe, Nevada)	83-71437	0-88318-199-1
No. 101	Positron-Electron Pairs in Astrophysics (Goddard Space Flight Center, 1983)	83-71926	0-88318-200-9
No. 102	Intense Medium Energy Sources of Strangeness (UC-Santa Cruz, 1983)	83-72261	0-88318-201-7
No. 103	Quantum Fluids and Solids - 1983 (Sanibel Island, Florida)	83-72440	0-88318-202-5
No. 104	Physics, Technology and the Nuclear Arms Race (APS Baltimore-1983)	83-72533	0-88318-203-3
No. 105	Physics of High Energy Particle Accelerators (SLAC Summer School, 1982)	83-72986	0-88318-304-8
No. 106	Predictability of Fluid Motions (La Jolla Institute, 1983)	83-73641	0-88318-305-6
No. 107	Physics and Chemistry of Porous Media (Schlumberger-Doll Research, 1983)	83-73640	0-88318-306-4
No. 108	The Time Projection Chamber (TRIUMF, Vancouver, 1983)	83-83445	0-88318-307-2
No. 109	Random Walks and Their Applications in the Physical and Biological Sciences (NBS/La Jolla Institute, 1982)	84-70208	0-88318-308-0
No. 110	Hadron Substructure in Nuclear Physics (Indiana University, 1983)	84-70165	0-88318-309-9
No. 111	Production and Neutralization of Negative Ions and Beams (3rd Int'l Symposium, Brookhaven, 1983)	84-70379	0-88318-310-2
No. 112	Particles and Fields-1983 (APS/DPF, Blacksburg, VA)	84-70378	0-88318-311-0
No. 113	Experimental Meson Spectroscopy - 1983 (Seventh International Conference, Brookhaven)	84-70910	0-88318-312-9
No. 114	Low Energy Tests of Conservation Laws in Particle Physics (Blacksburg, VA, 1983)	84-71157	0-88318-313-7
No. 115	High Energy Transients in Astrophysics (Santa Cruz, CA, 1983)	84-71205	0-88318-314-5
No. 116	Problems in Unification and Supergravity (La Jolla Institute, 1983)	84-71246	0-88318-315-3
No. 117	Polarized Proton Ion Sources (TRIUMF, Vancouver, 1983)	84-71235	0-88318-316-1

No. 118	Free Electron Generation of Extreme Ultraviolet Coherent Radiation (Brookhaven/OSA,1983)	84-71539	0-88318-317-X
No. 119	Laser Techniques in the Extreme Ultraviolet (OSA, Boulder, Colorado, 1984)	84-72128	0-88318-318-8
No. 120	Optical Effects in Amorphous Semiconductors (Snowbird, Utah, 1984)	84-72419	0-88318-319-6
No. 121	High Energy e^+e^- Interactions (Vanderbilt, 1984)	84-72632	0-88318-320-X
No. 122	The Physics of VLSI (Xerox, Palo Alto, 1984)	84-72729	0-88318-321-8
No. 123	Intersections Between Particle and Nuclear Physics (Steamboat Springs, 1984)		0-88318-322-6